Jinsong Zhou
Histochemistry
De Gruyter Graduate

Also of interest

Microbial Applications
Recent Advancements and Future Developments
Kumar Gupta, Zeilinger, Ferreira Filho, Carmen
Durán-Dominguez-de-Bazua, Purchase, (Eds.), 2016
ISBN 978-3-11-041220-8, e-ISBN 978-3-11-041282-6

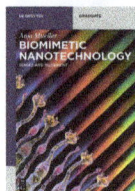

Biomimetic Nanotechnology
Senses and Movement
Mueller, 2017
IISBN 978-3-11-037914-3, e-ISBN 978-3-11-037916-7

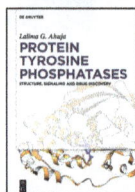

Protein Tyrosine Phosphatases
Structure, Signaling and Drug Discovery
Ahuja, 2018
ISBN 978-3-11-042643-4, e-ISBN 978-3-11-042177-4

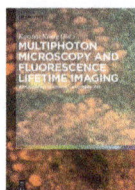

Multiphoton Microscopy and Fluorescence Lifetime Imaging
Applications in Biology and Medicine
König (Ed.), 2017
ISBN 978-3-11-043898-7, e-ISBN 978-3-11-042998-5

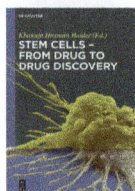

Stem Cells - From Drug to Drug Discovery
Haider, Husnain, 2017
ISBN 978-3-11-049628-4, e-ISBN 978-3-11-049376-4

Jinsong Zhou

Histochemistry

—

DE GRUYTER

西安交通大学出版社
XI'AN JIAOTONG UNIVERSITY PRESS

Author
Jinsong Zhou
Xi'an Jiaotong University
Xi'an, Shaanxi, China
zjs301@xjtu.edu.cn

ISBN 978-3-11-052482-6
e-ISBN (PDF) 978-3-11-053139-8
e-ISBN (EPUB) 978-3-11-053181-7

Library of Congress Cataloging-in-Publication Data
A CIP catalog record for this book has been applied for at the Library of Congress.

Bibliographic information published by the Deutsche Nationalbibliothek
The Deutsche Nationalbibliothek lists this publication in the Deutsche Nationalbibliografie;
detailed bibliographic data are available on the Internet at http://dnb.dnb.de.

© 2017 Walter de Gruyter GmbH, Berlin/Boston
Cover image: © Jinsong Zhou
Typesetting: Compuscript Ltd. Shannon, Ireland
Printing and binding: CPI books GmbH, Leck
∞ Printed on acid-free paper
Printed in Germany

www.degruyter.com

Preface

The composition of *Histochemistry* is under the supervision of highly experienced professors and technicians and is supported by De Gruyter Publishing House and Xi'an Jiaotong University Publishing House. The main purpose of this textbook is to cultivate the basic theory thinking and experimental skills and to enhance the research activity and spirit of innovations to match the needs of modern cultivation and education.

Histochemistry is a basic biomedical science and technology for biomedical post-graduate students and is widely used in life science research. To make sure the readers could thoroughly understand and master the basic theories and experiment methods, review questions or knowledge linkages are listed for each chapter, and experiment procedures and corresponding laboratory result images are added in this textbook. Because histochemistry is one of the fast-developing life sciences, the advanced methods in the fixation, the new fluorescence dyes such as cell markers, the use of flow cytometry, and the confocal laser scanning microscope and other new technologies are emphasized. How to save and analyze the experiment images is also a crucial question in this field; thus, the photomicrography is also introduced. This textbook can also be used as reference for other specialties for research purposes, such as basic medicine, preventive medicine, forensic medicine, oral medicine, nursing, pharmacy, and clinical medicine.

Jinsong Zhou, PhD
Xi'an Jiaotong University
March 2017

Acknowledgments

We thank Professor Tianbao Song at Xi'an Jiaotong University and Professor Zhen Li at Fourth Military Medical University who critically read the entire book and gave us important suggestions, and Ming Li at Xi'an Jiaotong University who gave us technological support. We also thank family members of contributors for their deep understanding of hard compiling work. We also extend our appreciation to the staff of Xi'an Jiaotong University Publishing House, Yuan Bao and Yuanyuan Wen, and staff of De Gruyter, Anne Hirschelmann, for their editorial enthusiasm and expertise.

Jinsong Zhou, PhD
Xi'an Jiaotong University
March 2017

Contents

Contributing authors

Jinsong Zhou, PhD: Chapters 1, 2, and 8
Associate Professor
Department of Human Anatomy, Histology and
Embryology
School of Basic Medical Sciences
Xi'an Jiaotong University Health Science Center
Xi'an, Shaanxi, PR China

**Xiaotian Zhang, PhD: Chapters 6, 15, and
Appendix II**
Assistant Professor
Department of Human Anatomy, Histology and
Embryology
School of Basic Medical Sciences
Xi'an Jiaotong University Health Science Center
Xi'an, Shaanxi, PR China

Ying Xu, BA: Appendix I
Associate Professor
School of Foreign studies
Xi'an Jiaotong University Health Science Center
Xi'an, Shaanxi, PR China

Bo Kou, PhD: Chapters 13 and 14
Assistant Researcher
Department of Cardiovascular Surgery
First affiliated hospital
Xi'an Jiaotong University
Xi'an, Shaanxi, PR China

Hong Tian, MS: Chapters 7, 9, and 10
Assistant Professor
Department of Human Anatomy, Histology and
Embryology
School of Basic Medical Sciences
Xi'an Jiaotong University Health Science Center
Xi'an, Shaanxi, PR China

Yuanjie Li, PhD: Chapters 3, 4, and 5
Assistant Professor
Department of Human Anatomy, Histology and
Embryology
School of Basic Medical Sciences
Xi'an Jiaotong University Health Science Center
Xi'an, Shaanxi, PR China

Ming Lu, MS: Chapters 11 and 12
Assistant Professor
Department of Human Anatomy, Histology and
Embryology
School of Basic Medical Sciences
Xi'an Jiaotong University Health Science Center
Xi'an, Shaanxi, PR China

Hongcheng Zhou, BE: Chapter 16
School of Information Engineering
Chang'an University
Xi'an, Shaanxi, PR China

1 Introduction

The development of life sciences requires deep understanding about the chemical compositions and their structural and functional relationships. During the past hundreds of years, more and more theories, technologies and methods were found and created to detect the chemical compositions in tissues and cells, and a borderline discipline, histochemistry, was established. As a methodology, nowadays histochemistry is widely used in basic life science studies and clinical application.

1.1 The History, contents and theories in histochemistry

1.1.1 Origin and development of histochemistry

Histochemistry is also known as microchemistry, which means the reaction processes and results will be observed not with naked eyes or in the test tube but under a microscope. At the beginning, the chemistry methods just occupied a little in histochemistry contents because of very few chemistry knowledge. In the 20th century, those eager to know the nucleic acids, the proteins and the enzymes require the theories and methods of histochemistry to develop very fast; thus, more and more new methods were created and verified. For example, the Feulgen reaction and the methylene green-pyronine were used to display the nucleic acid, PAS reaction was introduced to show glycogens and the calcium-cobalt and lead nitrate methods were applied to demonstrate the alkaline phosphatase and acid phosphatase. The immunohistochemistry method was created when specific antigen-antibody reaction was used to detect the antigen compositions in tissues and cells, and the molecular biology technologies, such as *in situ* hybridization and polymerase chain reaction (PCR), were introduced and combine with histochemistry. In the meantime, the quantitation assay technologies, such as photomicrography, image analysis, flow cytometry and laser scanning confocal microscopy, were established and taken in application.

Thus, as a newly established borderline discipline, histochemistry knowledge contains the information in histology and its related lasted methods, advanced technologies and discoveries. Thus, the concept of histochemistry should be like this: Based on histology, the modern methods and technologies in physics, chemistry, biochemistry, immunology and molecular biology are introduced to detect the chemical compositions *in situ*, and the qualitative and quantitative analysis are conducted to understand the normal and abnormal rules about the metabolisms, functions and morphology changes in cells and tissues.

DOI 10.1515/9783110531398-001

1.1.2 Contents of histochemistry studies

1.1.2.1 Inorganic materials in cells
These inorganic materials in cells and tissues mainly include the different kinds of metal ions, such as Ca^{2+}, Mg^{2+}, K^+, Zn^{2+}, Fe^{2+}, Fe^{3+}, Cu^{2+}, Pb^{2+}, Ag^+, Au^{3+}, As^{3+}, U^{6+} and I^-, and their salts, such as chloridate, phosphate, carbonate and nitrate.

1.1.2.2 Organic materials in cells
These organic materials in cells and tissues mainly include sugar (such as glycogen, starch, glycoprotein and proteoglycan), lipid (such as phospholipid, glycolipid, lipoprotein, lipoid, cholesterol and sterol ester), nucleic acid (such as DNA and RNA), peptide, protein, pigments and different kinds of vitamins.

1.1.2.3 Different kinds of enzymes
As special proteins, the enzymes play important roles in cell metabolism. Today, more than 200 kinds of enzymes can be displayed by histochemistry methods, such as acid phosphatase, alkaline phosphatase, 5-nucleotidase, glucose-6-phosphatase, adenosine triphosphatase, adenosine triphosphatase, carbonic anhydrase, nonspecific esterase, cholinesterase, cytochrome oxidase, peroxidase, monoamine oxidase, succinate dehydrogenase, lactate dehydrogenase, 3b-hydroxy steroid dehydrogenase, acylase and phosphorylase.

1.1.2.4 Antigen and antibody in cells and tissues
The specific antigen-antibody reaction can be used to detect not only the antigens in tissues and the antigen from pathogenic microorganism but also the autoantibody and antibody-antigen complex to provide reliable results for basic research and clinic diagnosis.

1.1.2.5 Endogenous and exogenous gene segments
With or without PCR technology, the *in situ* hybridization technology can be used to detect endogenous gene segments and their normal and abnormal expression, such as DNA, mRNA and the gene segments of virus, which can be used as references for clinical diagnosis.

1.1.3 Theories of histochemistry

1.1.3.1 Chemistry reaction
The already-known chemical reactions are used to demonstrate the materials in tissues and cells by forming colorful deposits or high electron density structures in one or several steps, such as the enzyme histochemistry. Alternatively, first, the

chemicals used for detection will be structurally changed *in situ* by the need-to-know materials to be suitable for subsequent reactions, and second, the additional chemical regents are applied to indirectly demonstrate the need-to-know materials in cells and tissues, such as PAS and Feulgen reaction.

1.1.3.2 Physics theory
The materials can be visualized by their physics characteristics, for example, the fat can be shown by colorful Sudan series dyes because the dyes are lipid-soluble, and the florescence can be observed when monoamine (such as noradrenaline, dopamine and 5-hydroxytryptamine) is induced by formaldehyde under the fluorescence microscope.

1.1.3.3 Biological characteristic
Most biomacromolecules can be regarded as antigens, which will be combined and visualized by the specific antibodies that are marked by fluorescence, enzymes or colloid gold, and this is the basic theory of the widely used immunohistochemistry. Some affinity chemical reactions, for example, the reaction between avidin and biotin, forms a branch of histochemistry, the affinity histochemistry.

1.1.3.4 Nucleotide chain complementary theory
The two complementary nucleotide chains will combine with each other to form stable hybrid. When one of them is labeled by a marker, it can be used in the *in situ* hybridization to show another (target) chain.

1.2 Basic requirements of histochemistry methods

The different theories, procedures and results match different histochemistry methods, but they all have the same purpose: to display the tissue and cell chemicals *in situ*. Thus, they must follow the basic common requirements.

1.2.1 Basic requirements

1.2.1.1 Specificity
The high specificity of reactions against the need-to-know materials guarantees the right experiment results.

1.2.1.2 Sensitivity
The high sensitivity of reactions makes sure that the method can be used to detect trace need-to-know materials in cell and tissue.

1.2.1.3 Fixation

The perfect fixation will provide with almost an authentic cell, tissue and material structures before histochemistry staining, which is necessary for *in situ* observation and records.

1.2.1.4 Reaction deposits

The reaction deposits must be formed *in situ*, insoluble, stable and colorful for light microscopy or with high electron density for electronic microscopy.

1.2.1.5 Repeatability

The repeatability is a basic rule for all kinds of scientific studies.

1.2.2 Things you need to know

a. The characteristics of need-to-know materials, such as water solubility, lipid solubility and the possible location in cell and tissue, must be clear before experiment. This means many references must be thoroughly read and understood.
b. The conditions of each procedure in experiments must be strictly controlled, such as the concentration of each kind of chemical reagent, the temperature and the pH value of reaction liquor, especially for the enzymes. In the control test and following repeated experiments, the conditions should remain the same.
c. The control tests must be set up along the experiments. The positive control test is used to demonstrate the experiment method, procedures and reagents are effective, and the negative test is used to demonstrate the experiment results are specific, which is crucial for result analysis.
d. The reagents in use should be analytically pure (AR class) and will not influence the chemical characteristics of need-to-know materials or the enzyme activity.
e. All the laboratory utensils must be clean and unpolluted. The water in use should be double distilled water (DD H_2O).

Review question

Generally say, what are the differences of the study fields among histology, histochemistry and immunohistochemistry?

2 Tissue preparation

In general, the specimen could be studied with histochemistry and immunohistochemistry only after the processes of tissue collection, fixation, embedding and sectioning.

2.1 Tissue collection

2.1.1 Attention

a. Keep the instruments sharp and clean.
b. Collect the sample quickly and accurately.
c. Keep the sample away from artificial damage.
d. It's better to collect the tissue at low temperature ($0°C–4°C$).

2.1.2 Size of specimen

The specimen should be kept small and thin because the fixatives penetrate through the sample thoroughly. In general, the size of specimen for light microscope study is approximately $1 \times 1 \times 0.5$ cm and less than 1 mm^3 for most electron microscope study.

2.2 Fixation

Fixation means that the specimen is treated with one kind or a mixture of different kinds of chemicals. The fixative needs to permanently preserve the tissue structure and the chemical compositions just like when alive. Specimens should be fixed immediately after they are removed from the body.

2.2.1 Purpose of fixation

Fixation preserves the tissue and cell in living condition as much as possible for subsequent treatment. Fixation is used
a. to inhibit autolysis – after the tissue is isolated from body, the lysosome structure breaks and releases lysosomal enzymes because of oxygen deficiency, which results in cell damage. This process is known as autolysis. The fixative can inactivate lysosomal enzymes.

DOI 10.1515/9783110531398-002

b. to prevent solution – tissue compositions are dissolved during tissue preparation because most reagents can dissolve chemicals in tissues. The chemicals in the fixative react and precipitate most large molecules in tissue.
c. to avoid corruption – fixatives fix proteins to kill bacteria.
d. to reduce injury – fixation makes the cytoskeletal proteins more stable than before.
e. to harden the tissue – fixatives harden the soft tissue, which is beneficial for sectioning.
f. to alter refractive index – fixation changes the refractive index of cell and tissue composition to some extent, which makes the structures to be discerned easier.
g. to enhance the dyeing – in a sense, some fixatives will promote dyeing effect for some reasons.

2.2.2 Object for fixation

The fixatives mainly react with proteins because they take functions probably in the following way:
a. The chemical bonds between fixative and protein form the precipitates.
b. Fixatives interrupt the degeneration of protein. Protein will lose solubility after degeneration.

Basically, other materials such as lipid, sugar and sugar-like chemicals, nucleic acid and other large molecules are combined with different kinds of proteins, and they will be fixed by protein-fixation processes.

2.2.3 Quality of the fixative

The fixatives usually should possess the following qualities:
a. Fixatives penetrate the tissue and cell quickly but do not change the tissue structure.
b. Fixatives produce tissue swelling and contraction as little as possible.
c. Specimen could be saved in the fixatives for a relatively long period.

2.2.4 Methods for tissue fixation

2.2.4.1 Soak fixation
Soak fixation is also called immersion fixation. In this kind of fixation, the following steps are followed: keep the fixation in 0°C–4°C, place several layers of gauze and absorbent cotton on the bottom of glass container and then put the specimen in the fixative for several hours at least.

2.2.4.2 Perfusion fixation

The fixative is injected through the whole body or an organ by cardiovascular system. In this way, the living cells are fixed *in situ* quickly. Buffer or normal saline containing proper amount of heparin is used to treat the blood before the fixatives are injected into the cardiovascular system. The amount of fixatives and the perfusion pressure are determined according to different animals. Tissue is collected after perfusion fixation and immersed into the same fixative if necessary.

2.2.4.3 Cultured cells fixation and smear fixation

As for monolayer culture, remove the coverslip from the culture plate and perform fixation. As for suspension culture, prepare smear after concentration by centrifugation. Then the smear is fixed in fixatives. Usually, the glass slide needs to be coated by adherence chemicals before using to prevent tissue slice shedding.

2.2.5 Notes for fixation

a. Fixed specimen must be fresh. The specimen should be fixed immediately after removal from the animals. Long time delay will result in the volume shrinkage or autolysis, and this is the big taboo of histochemistry and immunohistochemistry.
b. Fixed specimen should be put into sufficient quantity of the fixative. The general fixation dose is from 20 to 50 times the size of tissue block because fewer fixatives will result in insufficient fixation due to the influence of tissue fluid; on the other hand, too much fixative is not necessary.
c. Fixation duration time depends on the tissue type and size and the kind of fixatives. Generally, it is 24 hours for tissue block and maybe 15 seconds for cultured cells in a single layer on the glass slide.
d. In general, the specimen can be fixed at room temperature. The fixative for enzyme detection should be placed in the refrigerator before using.
e. The appropriate fixative is selected depending on the different types of tissue. Before histochemistry and immunohistochemistry experiments, the selection of fixatives is tricky, especially for new tissue or new materials that will be detected. In this case, reading more references and doing more preexperiments are strongly recommended.

2.2.6 Frequently used fixative

Different fixatives are usually divided into the following four types according to their chemical characteristics:

Type 1: aldehyde, such as formaldehyde, glutaraldehyde, paraformaldehyde, acrylaldehyde, malondialdehyde and so on

Type 2: oxidant, such as osmic acid, kalium hypermanganicum (KMnO$_4$), bichromicum kalium and so on

Type 3: protein denaturation agent, such as methyl alcohol, alcohol, acetic acid and so on

Type 4: others, such as mercuric chloride, picric acid and so on

According to the usage methods, the fixative can be used solo or with other fixative(s) as a mixture. Thus, they can be divided into pure and mixture fixatives.

2.2.6.1 Pure fixative

Commonly used pure fixative includes formaldehyde, alcohol, glacial acetic acid, picric acid, bichromicum kalium, osmic acid, mercuric chloride and acetone. Formalin, alcohol and acetone are commonly used as pure fixatives; the others are used as an ingredient of mixed fixatives.

2.2.6.1.1 Formaldehyde

Formaldehyde is a gas, and its saturated water solution contains 37%–40% formaldehyde and is called formalin. Generally, 10% formalin (4% formaldehyde solution) is used for a fixative. Formaldehyde is also as stable as the solid form called paraformaldehyde, the high molecular weight polymer.

Formaldehyde is a reducing agent. In water, formaldehyde monomer is a monohydrate, namely, methylene glycol (CH$_2$(OH)$_2$). It is the most effective fixative. Formaldehyde connects the adjacent proteins with bridging bonds and becomes insoluble polymer. Common formaldehyde reaction is that it is added to a reactive hydrogen compounds and forms methylol compounds. Any further condensation forms methylene bridges with other hydrogen atoms. The equation is shown in Fig. 2.1.

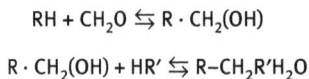

$$RH + CH_2O \leftrightarrows R \cdot CH_2(OH)$$

$$R \cdot CH_2(OH) + HR' \leftrightarrows R{-}CH_2R'H_2O$$

Fig. 2.1: The fixative mechanism of formaldehyde.

Formaldehyde has many advantages as fixative. It has strong penetrability, can save most of material compositions in the original state in tissue and can even fix bigger tissue block. However, it hardens tissue, shrinks tissue slightly and increases the flexibility of specimen. Formaldehyde not only fixes proteins but also saves the lipid and glycogen very well. At last, formaldehyde is also a good specimen preservation solution. However, formaldehyde has also some disadvantages. Formaldehyde contains more impurities, such as methanol, which decreases enzyme activity. Formaldehyde contains traces of formic acid, which can change the pH level of fixative and influence the dyeing effect. Formaldehyde is volatile and makes the specimen to be dried up. Formaldehyde needs long time to be removed by running water, and if not clearly removed, the dyeing effect will be greatly influenced. Formaldehyde can seal the antigen; thus, enzyme digestion is usually needed for antigen repair in immunohistochemical study when the tissue is fixed by formaldehyde.

The commonly used formaldehyde fixatives are as follows:

a. 10% buffered formalin fixative

Formaldehyde	100 mL
Distilled water	900 mL
NaH_2PO_4	4 g
Na_2HPO_4	6.5 g

This fixative has a good fixation effect with very little damage to tissue. The fixing agent is also suitable for immunohistochemical staining. This solution has a neutral pH level.

a. Salt 10% formalin fixative

Formaldehyde	100 mL
Distilled water	900 mL
NaCl	9 g

When preparing this fixative, NaCl and then formaldehyde are added into water. This fixative can promote and improve dyeing and increase the intensity of the staining because of NaCl.

a. Aldose calcium fixative

Formaldehyde	100 mL
Sucrose	300 mL
Calcium acetate	20 g
Distilled water	90 mL

When preparing this fixative, sucrose and calcium acetate and then formaldehyde are added to distilled water. The fixative is better for histochemical enzyme study. The specimen could undergo frozen section after fixation because the high concentration of sucrose in the fixative solution can prevent the formation of ice crystal, which could damage tissue and cell microstructure.

2.2.6.1.2 Ethanol, methanol and acetone

These three kinds of chemicals could be used to replace water component in protein structure so that the hydrogen bonds rupture, the protein tertiary structures are destroyed and the soluble proteins in the cytoplasm are coagulated.

Ethanol, also known as alcohol, is a colorless and transparent liquid. The high concentration of ethanol can be used to save glycogens and other substances. Ethanol has both fixation and dehydration effects. Ethanol cannot fix cells or tissue together with other oxidants such as chromic acid, potassium dichromate and osmic acid because ethanol itself is chemically reductant.

Acetone and ethanol can combine fixation and dehydration in one step. They can quickly make section and preserve the activity of enzyme in a certain degree. Even after paraffin embedding, enzyme activity is still saved in a sense. Therefore, ethanol has been used as one of the conventional fixatives. If the specimen is fixed in cold acetone after fresh frozen section or cryostat section, the activity of enzyme could be preserved better. If the specimen is fixed at low temperature ($-70°C$), the lipid can also be saved although it is soluble in acetone. Then the lipid can be shown by Sudan dyes. After formaldehyde and glutaraldehyde fixation, some lipids, such as phospholipids, will be no longer dissolved in fat solvent (acetone).

2.2.6.1.3 Glutaraldehyde

Glutaraldehyde, as a bifunctional aldehyde fixative, is often used in electron microscope technology. Glutaraldehyde is a good fixative to maintain glycogens, glycoproteins, microtubules, endoplasmic reticulum, cell matrix structure and enzyme activity. The disadvantages of glutaraldehyde are as follows: it cannot save fat, it has no electronic dyeing effect and it has lower power to display cell membrane.

The pH level of glutaraldehyde solution is 4.0–5.0. If the storage time is too long, the pH level of glutaraldehyde becomes 3.5 or even lower, and the fixation effect is reduced greatly. Commodity glutaraldehyde contains a certain amount of impurities. If it is used in cytochemistry, immunohistochemistry and other special experiments, glutaraldehyde should be used after purification.

The concentration of glutaraldehyde is 1%–3% for electron microscopy technology. Generally, phosphate or dimethyl arsenate buffer is used to prepare the glutaraldehyde fixative. Glutaraldehyde monomer becomes methyl glycol base ($-CH_2(OH)_2$) after hydrating and takes charge in the function of the fixative. The aldehyde group mainly reacts with the amino group of proteins, peptides and amino acid. This cross-linking is much more stable than the formaldehyde cross-linking. The reaction equation is shown in Fig. 2.2.

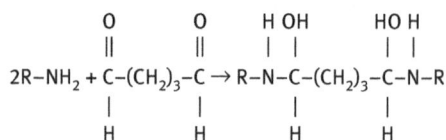

$$2R-NH_2 + \overset{\displaystyle O}{\overset{\|}{C}} - (CH_2)_3 - \overset{\displaystyle O}{\overset{\|}{C}} \rightarrow R - N - \overset{\overset{\displaystyle H}{|}}{\underset{\underset{\displaystyle H}{|}}{C}} - (CH_2)_3 - \overset{\overset{\displaystyle OH}{|}}{\underset{\underset{\displaystyle H}{|}}{C}} - N - R$$

Fig. 2.2: The fixative mechanism of glutaraldehyde.

2.2.6.1.4 Osmium tetroxide

Osmium tetroxide is a light yellow crystal with high toxicity and strong volatility. Osmium tetroxide is a strong oxidizer so it cannot be mixed with reductant such as alcohol and formaldehyde. The aqueous solution of osmium tetroxide is easily deoxidized into black precipitation of hydroxide osmium. The osmium tetroxide must be stored in the colored bottles because the natural light will facilitate the reduction. Usually, 10 drops of 5% mercuric chloride are added into 100 mL water solution to prevent the reduction.

Osmium tetroxide is a good choice to fix lipid because it reacts with unsaturated fatty acids. The osmium tetroxide fixes unsaturated fat and lipid in black as follows (Fig. 2.3).

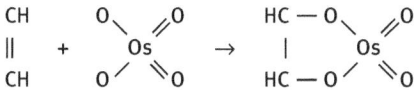

$$\begin{array}{ccc}
CH & O\diagdown\diagup O & HC-O\diagdown\diagup O \\
\parallel \quad + & Os & \rightarrow \quad | \qquad Os \\
CH & O\diagup\diagdown O & HC-O\diagup\diagdown O
\end{array}$$

Fig. 2.3: The fixation mechanism of osmium tetroxide for unsaturated lipid.

In the electron microscope technology, osmium tetroxide is a kind of important fixative, especially for phosphatide protein of cytoskeleton. Osmium tetroxide has strong electronic dyeing effect. The disadvantage of osmium tetroxide is that its penetrating speed is slow, but the fixation time should not be too long. In the electron microscope technology, osmium tetroxide is used as after-fixing fixative after formaldehyde and glutaraldehyde fixation.

2.2.6.1.5 Acetic acid
The pure acetic acid is also called glacial acetic acid. Acetic acid can be mixed with water or alcohol in various proportions. Acetic acid can precipitate nucleoprotein and has a quickly penetrating rate through the tissue. Thus, it is a good fixative for nucleus. However, acetic acid can make collagen fiber expansion. Acetic acid in the mixed fixative can resist organize shrinking.

2.2.6.1.6 Picric acid
Picric acid, or trinitrophenol, is a toxic yellow crystal. Dry picric acid triggers autoignition and even explosion in air. Thus, it should be stored in saturated water solution (yellow). The tissue becomes yellow after fixed in picric acid. The yellow color will fade by soaking in water or alcohol for a long time.

Picric acid is soluble in alcohol, dimethyl benzene, benzene and water. It can precipitate all kinds of proteins, forming water-insoluble picric acid-protein compounds (picrate). However, picric acid has no fixation function for fat and lipids. Picric acid has a slow penetration rate, makes the tissue significant contraction but does not harden the tissue.

2.2.6.2 Mixed fixative
Pure fixatives focus on a certain part or chemical component of the fixed cells or tissues. If a variety of ingredients in cells and tissues need to be fixed at the same time, the mixed fixatives prepared with multiple chemicals must be introduced. The

different chemicals in the mixed fixatives need to complement and cooperate with one another very well to gain the perfect results of fixation.

2.2.6.2.1 Zenker liquid
In Zenker liquid, 2.5 g potassium dichromate and 5 g mercuric chloride are dissolved in 100 mL heated (40°C–50°C) distilled water. Filter the liquid after cooling, and preserve it in a dark place to form a stock solution. When using the liquid, 5 mL glacial acetic acid is added into 95 mL stock solution and then used immediately. Both the potassium dichromate and the mercuric chloride are strong oxidizing agents. Zenker liquid is one kind of commonly used fixatives in histological and cytological research. After fixation in Zenker liquid, the staining of cell nuclear and cytoplasm is clear and stable. Generally, specimen is fixed for 24 hours, rinsed in running water for 24 hours and then dehydrated.

2.2.6.2.2 Carnoy liquid
Anhydrous ethanol, chloroform and acetic acid are mixed before using at the ratio of 6:3:1. The liquid possesses good penetrability, fixed power and nuclear fixed ability, and it saves glycogen very well. Anhydrous ethanol in the mixed fixative can fix cytoplasm and glycogen. Acetic acid can fix chromatin, prevent the hardening and contraction effect of alcohol and increase the penetrability. Moreover, chloroform can increase penetration. In view of the penetrating power of this liquid, small tissue block is fixed for 1–2 hours and regular size tissue block for 4–6 hours. After fixation without rinsing, specimen can be immersed into the 95% alcohol or directly into the anhydrous alcohol for dehydration. This liquid is applicable to general cytological studies (such as making specimens of cell division, showing DNA, RNA and liver glycogen). This mixed fixative possesses good effects to lymphoid tissue and glands and is suitable for almost all kinds of dyeing.

2.2.6.2.3 Bouin liquid
Picric acid saturated aqueous solution, formaldehyde and glacial acetic acid are mixed before using according to the ratio of 15:5:1. Picric acid precipitates protein with slow penetration speed and shrinks tissue. Glacial acetic acid and formaldehyde have quickly penetrating speed. Glacial acetic acid can make tissue expansion. Therefore, the three ingredients in the liquid cooperate with one another. Bouin liquid becomes a good fixative in many aspects. After fixation, the specimen can be treated directly into the 70% alcohol for dehydration, and the yellow color of the specimen fades.

2.2.6.2.4 Paraformaldehyde-glutaraldehyde fixative
10% paraformaldehyde	10 mL
0.2 mol/L phosphate buffer	25 mL
25% glutaraldehyde solution	5 mL
Double distilled water	10 mL

The pH level needs to be adjusted to 7.4 after mixture. The penetrating power of paraformaldehyde is better than that of glutaraldehyde. Moreover, the ability of paraformaldehyde to preserve enzyme activity is better than that of glutaraldehyde. However, glutaraldehyde can save the ultrastructure of cell much better. Therefore, paraformaldehyde-glutaraldehyde fixative cannot only save the ultrastructure and enzyme activity but can also be used in electron microscope cytochemistry technique.

2.2.6.2.5 Alcohol-acetone solution
Equal dose of anhydrous alcohol and acetone is mixed together. Acetone is a kind of protein precipitation agent, and it keeps the enzyme reactive groups in original conditions to a great extent. Thus, cold acetone can preserve enzyme. Pure alcohol has the ability to fix, harden and dehydrate the cells and tissue. The fluid is used in the fixation of hydrolytic enzymes.

2.2.6.2.6 Gendre liquid
Picric acid (90% saturated alcohol liquid)	80 mL
40% formaldehyde	15 mL
Glacial acetic acid	5 mL

If the mixture is precooled at −78°C before using, it will act as a substitute of freezing liquid. Alternatively, the specimen is fixed at 4°C for 6–8 hours. Gendre liquid fixes glycogen better than Bouin liquid.

2.2.6.3 How to choose fixative
The choice of fixative must be based on the properties of various substances. For example, formaldehyde is a good fixative for a variety of tissue or cells, and the ability to preserve lipid and enzyme is superior to other fixative. Glutaraldehyde and osmium tetroxide can preserve the cell ultrastructure very well. They are applied to electron microscopy techniques. Acetone and ethanol have good effect to save polysaccharide and protein. The acid Carnoy fixative is suitable for the fixation of RNA.

Pay attention to the balance of various reagents when choosing a mixed fixative. For example, reagent causing tissue contraction usually matches the reagent causing tissue expansion. The less penetration reagent can be mixed with the high penetrability reagent. However, the oxidation reagents cannot be mixed with reduction reagents.

To save more chemical and enzyme activity, fresh samples without fixation and the frozen section are strongly recommended. In this case, the section can be fixed after incubation or chromogenic reaction. On the other hand, the specimen can be slightly fixed after the frozen section and then undergo incubation and color processing, but this may cause color diffusion and needs proper control tests.

2.3 Tissue rinse, dehydration and clearance

2.3.1 Rinse

The fixative permeated into the specimen must be cleaned up thoroughly after fixation. Otherwise, the remaining fixative will influence the effects of dehydration and dyeing. Moreover, the remaining fixative could result in precipitation and crystallization in tissue, which will influence observation. In this case, the following needs to be noticed:

a. Do not require rinsing if the fixative is alcohol or alcohol mixture.
b. The fixative prepared with water should be clean up by flowing water.
c. If the fixative is oxidant, such as OsO_4, potassium dichromate, chromic acid and chromate, the specimen should be clean up by flowing water.
d. Specimen fixed by the fixative containing mercury should be rinsed by water or alcohol. Then the mercury is removed by iodine.
e. The fixative containing picric acid should be rinsed with 70% alcohol or water.
f. The rinse time depends on the kind of fixative, generally is 10–24 hours.

2.3.2 Dehydration

Specimen contains a large amount of water after fixation and rinse. Water cannot mix with transparent or embedding agents and hinder them to permeate the specimen. Therefore, dehydration is introduced to remove the water within the specimen. Dehydration is advantageous to transparence, wax immersion and saving the specimen permanently.

The dehydrating agent must be able to mix with water at any ratio. If the dehydrating agent can be paraffin soluble at the same time, such as N-butanol, the specimen can be directly engaged in wax immersion after dehydration. If the dehydrating agent, such as alcohol and acetone, is not soluble in paraffin, the specimen needs to undergo clearance by xylene before treated with paraffin.

Generally, the dehydrating agent should be prepared into various concentrations. Specimen should be treated by the dehydrating agent from low concentration to high concentration step by step, which gradually reduces the moisture in tissue and does not cause strong contraction or deformation.

2.3.3 Clearing

Most of the dehydrating agent cannot mix with paraffin. Thus, a reagent that can mix with both dehydrating agent and wax is needed. The reagent gradually penetrates into the tissue to exclude the dehydrating agent. When the dehydrating agent is replaced completely, light can pass through the specimen. The specimen shows varying degrees of transparent state, known as cleaning. The reagent is called clearing agent. The purpose of clearing is convenient for paraffin embedding.

All the clearing agents, such as xylene, toluene, benzene, chloroform, wintergreen oil and cedar oil, are soluble in paraffin. The commonly used clearing agents are xylene and chloroform.

2.4 Embedding

After dehydration and clearance, the paraffin for light microscopy technology or the epoxy for electron microscopy technology penetrates into the sample inside to support and harden specimen for sectioning. This process is known as embedding.

An excellent embedding medium must own the following properties: (1) the embedding medium is easy to change from a liquid status to a solid status after soaking the tissue completely; (2) the liquid embedding medium is easy to penetrate into the tissue, does not react with tissue composition and does not extract and dissolve the cell components; (3) the volume of specimen has no obvious change once the embedding medium becomes solid state; (4) the embedding medium has good cutting properties, such as homogeneous, strong, plasticity and elasticity; and (5) before dyeing, the embedding medium is easy to be removed or do not impede the dye to penetrate specimen.

2.4.1 Embedding medium for light microscopy technology

The ideal embedding medium for light microscopy is paraffin. Paraffin is a translucent crystal block at room temperature. Paraffin is divided into soft and hard wax depending on the melting point. Melting point for soft wax is 52°C–56°C and for hard wax is 60°C–62°C. Hard wax is used at higher room temperature, and soft wax is used at lower temperature. Hard and soft waxes are used to cut thin and thick section, respectively. In the embedding procedures, the specimen is immersed in paraffin I (paraffin and xylene mixture), paraffin II and then paraffin III. The key point for paraffin embedding is the temperature.

2.4.2 Embedding medium for electron microscopy technology

The ideal embedding medium for electron microscopy should possess the following properties:
a. It is easy for the medium to penetrate the tissue.
b. The medium should be homogeneous after polymerization.
c. The volume of medium becomes smaller after polymerization.
d. The medium is tolerant for electron bombardment.

e. The medium has no deformation under the high temperature.
f. The medium keeps fine structure in good condition.
g. The embedding medium has no visible structure under electron microscopy.
h. The medium has a good cutting performance.

Epoxy resin is the polymer of epichlorohydrin and polyhydroxy compounds. Its structure is shown in Fig. 2.4.

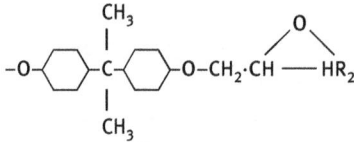

Fig. 2.4: The structure of epoxy resin.

Epoxy resin is a thermoplastic resin. Epoxy resin itself cannot be aggregated into pieces. It must be mixed with a certain amount of curing agent and other auxiliary agent to form the formation of irreversible yellow brown solid under the high temperature.

2.5 Sectioning

Tissue section performance in histochemistry usually includes paraffin sectioning, frozen sectioning and vibration sectioning method.

2.5.1 Paraffin section method

The paraffin section method is the better way to save time, and it is easy to operate to gain very thin slices of tissue (usually 6–10 μm thick) and also easy to make serial section. The tissue block embedded in paraffin can be stored permanently. However, the paraffin section method may change the chemical reactions inside the tissue, lose the enzymes and other substances (such as lipid) completely, make protein denaturation and inactivate enzyme activity.

2.5.2 Frozen section method

The frozen section method is the most commonly used technology in enzyme histochemistry. The fixed or unfixed tissue samples can be treated by frozen section. Frozen section for fresh tissue keeps enzyme activity in tissues or cells very well, especially for the enzymes and antigens that are highly sensitive to heat, weak acid, alkali, organic solvents, etc.

In frozen section, water in the tissue is easy to form ice crystals, which affects the location of enzyme and antigen. The following methods are used to reduce the formation of ice crystals:

1. Shock cooling. Shock cooling means the fresh or fixed sample undergoes deep cooling in a very short time, for example, in a few seconds. This method can reduce the ice crystal formation and requires that the tissue block is smaller and under the OCT (optimum cutting temperature compound, a mixture of polyethylene glycol and vinol in water solution) protection. OCT is a special gel-like liquor that can protect the tissue block under the deep frozen temperature. After the specimen is immersed in OCT embedding agent, it will be treated buy liquid nitrogen for conservation.

2. Freeze protection. The freeze protection materials are used to prevent samples from frost damage and to keep the activity of cells and tissues in freezing condition. These materials include glycerin, dimethyl sulfoxide, sucrose, mannose and polyvinylpyrrolidone. Now, the specimen is stored in cold glue-sucrose liquid fixation in cold formalin. The specific steps are as follows:

   a. The specimen is fixed in the following liquid for 18–24 hours at 0°C–4°C.

40% formaldehyde	10 mL
Phosphate buffer	50 mL
Sucrose	7.5 g
Add distilled water to 100 mL	

   After fixation, the tissue block is slightly rinsed with distilled water at 0°C–4°C.

   b. The specimen is treated with the following liquid at least 12–24 hours at 0°C–4°C.

Sucrose	30 g
Arabic gum	1 g

   Mix them and make sure the gel is formed without precipitation. Then add 100 mL and one grain thymol.

   c. Cryostat frozen section. Cut the tissue block into approximately 3-μm-thick slices continuously under –30°C low temperature. Generally, the fixed or unfixed tissue block is cut at very low temperature after special protection treatment against frost, and then the sections undergo other operation procedures. Because some chemicals in tissue such as enzymes will be damaged or cannot be well preserved in the fixation, this sectioning method is a good choice.

2.6 Adherence and mounting

2.6.1 Adherence and adhesive medium

Generally, the flattened fresh tissue slice can adhere to clean the glass slide directly, and the adhesive medium is unnecessary. However, in most cases, adhesive medium is used to make sure that the tissue slice is attached to the glass slide tightly because the slice needs to undergo different kinds of treatments with different chemical solutions under different temperature levels.

2.6.1.1 Protein-glycerin mixture

It is the most commonly used adhesive medium for histology and some histochemistry and immunohistochemistry experiments. It is also easy to prepare: Add egg white to same amount of glycerin and mix them well. The lower clear solution is collected after stewing, then add a small amount of thymol as antiseptic. This adhesive medium is also cheap, but because the chemical compositions of egg white are still unknown clearly, it may affect the staining result, for example, the biotin in egg white will combine with avidin so the ABC methods in immunohistochemistry are not recommended when the slice is attached to the glass slide by protein-glycerin mixture.

2.6.1.2 Chrome alum gelatin

The advantages of chrome alum gelatin include that it adheres to slide firmly, does not influence the dyeing and does not to be affected by acid and alkali circumstances. To prepare this adherence medium, 5–10 g gelatin and 0.5 g chrome alum need to be dissolved in 1000 mL distilled water. First, put the wet gelatin into incubator to stay overnight; second, add water and heat; and finally, add chrome alum. Cool and filter the liquor, and add a small amount of thymol and store in the refrigerator.

2.6.1.3 Poly-L-lysine

Add 10 mg poly-L-lysine into 100 mL double distilled water to form the poly-L-lysine adhesive medium. When using, put a drop of poly-L-lysine liquid on the slide, push the liquid into a film and dry in the air. The poly-L-lysine forms positive charge on slide surface, which combines with negative charge of specimen slice. Therefore, the tissue slice is firmly adhered to the glass slide treated with poly-L-lysine.

2.6.2 Mounting and mounting medium

Mounting medium can seal the slice and save the section forever, which facilitates to microscopic examination. Mounting medium should possess the following properties: the counting medium can mix with clearing medium very well, the counting medium does not affect dyeing effects, the mounting medium has similar refractive index with the glass slide and the mounting medium has some stickiness.

The mounting medium is generally divided into wet and dry mounting medium.
a. Wet mounting medium
    If the slice will be mounted with wet mounting medium, it does not need dehydration and clearing treatments.
    i.   Pure glycerin. Pure glycerin is generally used to seal motor end plates for nerve tissue, or the separation specimen of epithelial tissue and adipose tissue.
    ii. Glycerin gelatin. To prepare this mounting medium, the gelatin is added to distilled water and dissolved by heating. Add the appropriate amount of glycerin and keep in the refrigerator. The refractive index of glycerin gelatin is higher than that of the glycerol, and glycerin gelatin has certain hardness. To use the medium, dissolve it by heating.
b. Dry mounting medium
    The section must be treated with alcohol dehydration and xylene clearing.
    i.   Canada balsam. Canada balsam, as the common mounting medium, is a transparent light yellow liquid. Canada balsam is soluble in xylene and is volatile and dry faster.
    ii. Neutral balsam. Add anhydrous sodium carbonate into xylene gum, and keep stirring in a low-temperature refrigerator. After a few days, the supernatant will be collected as the mounting medium.

2.7 Buffer

2.7.1 Composition and application of buffer

Many histochemical reactions depend on the pH level of the reaction system. In the process of histochemistry staining, the required pH level is crucial because the enzymes in the processes of staining need its special and stable pH level.

The buffer fights against a small amount of strong acid and alkali from outside and causes less pH change of solution. If a solution has such an effect, it is known as buffer. Usually in the buffer, there is at least one of the following pairs of components:
a. The weak acid and its salt, for example, HAC and NaAC.
b. The acid salt of polybasic acid and its corresponding secondary salt, for example, $NaHCO_3$ and Na_2CO_3.
c. The weak base and its salt, for example, NH_3 and NH_4Cl.

Here, the acetic acid-sodium acetate buffer system is used as an example to illustrate the functional mechanism of buffer. Acetate is a weak electrolyte, which partially ionize into hydrogen ions and acetate ions in solution. Sodium acetate is a strong electrolyte, which will ionize completely into sodium ions and acetic acid ions in solution (Fig. 2.5).

$$HAC \leftrightarrows H^+ + AC^-$$

$$NaAC \leftrightarrows Na^+ + AC^-$$

Fig. 2.5: The functional mechanism of buffer.

Before the buffer functions, the ions in the buffer include H^+, AC^-, Na and OH^- that come from the ionization of water molecule itself. All the ions are stable, and their concentration is balanced; thus, the buffer has its own pH level. When a small amount of acid or alkaline substances is ran into the reaction system by accident, they can be neutralized by OH^- or H^+, respectively, in the buffer, and the concentration of all kinds of ions will be changed accordingly and rebalanced again. In this way, the buffer system can make the pH level stable. Generally, if the concentrations of chemicals are higher, the buffer will have a stronger ability to keep the pH level stable, but the high concentration of chemicals also exhibits high ionic strength, which may influence some histochemistry and immunohistochemistry reactions.

2.7.2 Commonly used buffer

For the commonly used buffer, see Appendix I for details.

Knowledge links

Agar NY, Yang HW, Carroll RS, et al. Matrix solution fixation: histology-compatible tissue preparation for MALDI mass spectrometry imaging. Anal Chem. 2007 79(19):7416–23.

3 Carbohydrate and its derivatives in histochemistry

Carbohydrates are one of the major organic components of biological tissues. As the stored forms of energy, they also combine with other macromolecules to form the intracellular or extracellular various structures.

Carbohydrate histochemistry is a technology to display the sugar by proving some reactive groups of sugars in tissue sections. Free carbohydrates are not easily demonstrated with most histochemical technology because of its water solubility, but macromolecular polymer, such as glycogen, glycoproteins, proteoglycans, etc., can be displayed by histochemical technology.

3.1 Classification

3.1.1 Chemistry of carbohydrates

Carbohydrates are the polyhydroxy compounds, which can be divided into the following categories.

3.1.1.1 Monosaccharides
Monosaccharides are the basic units of oligosaccharides and polysaccharides. A large number of monosaccharides are five-carbon simple carbohydrate (pentaglucose) and six-carbon simple carbohydrate (hexose) in animal tissues. Among the hexose, the most important is glucose, its second and third carbon atoms conjugate hydroxyl to form the structure called glycol group. In addition to glucose, which already exists in the form of monosaccharides, other hexoses joined by glycosidic bonds form oligosaccharide or polysaccharide. They are called sugar residues in oligosaccharide and polysaccharide chains.

3.1.1.2 Oligosaccharides
Oligosaccharide is made up of less than 20 sugar residues. Oligosaccharide chains and their covalent proteins or lipids form the glycoproteins or glycolipids.

3.1.1.3 Polysaccharides
Polysaccharides contain more than 20 sugar residues. It can be classified into homopolysaccharide and heteropolysaccharide. When all the monosaccharides in a polysaccharide are the same type, these polysaccharides are known as homopolysaccharides, and they are called heteropolysaccharides when more than one type of monosaccharide is present.

DOI 10.1515/9783110531398-003

Glycogen is the only homopolysaccharide type found in animals that is composed of glucose polymerization. Heteropolysaccharides are long unbranched polysaccharides consisting of the same repeating disaccharide units. The repeating unit consists of an amino sugar along with uronic acid; therefore, the heteropolysaccharide is also called glycosaminoglycan (GAG).

3.1.1.4 Glycoproteins

Glycoproteins are conjugated proteins that contain a small amount of oligosaccharide chains covalently attached to polypeptide side chains. Generally, based on its distribution and function, it can be divided into following classifications: mucous glycoprotein, seroglycoid, structural glycoproteins and membrane glycoproteins.

3.1.1.5 Proteoglycans

Proteoglycans are composed of GAGs as the main structural components combined with a small amount of proteins, and its molecule consists of the following: hyaluronan-binding protein (core protein·[GAG] n) (Fig. 3.1). The same subunit of proteoglycan often contains more than one GAG side chains. Because various polysaccharides in the GAGs contain many polyanion, such as hydroxyl, carboxyl and sulfate groups, the GAGs are also called polyanionic aminohexoses. These anions are structural basis of carbohydrate histochemical reaction.

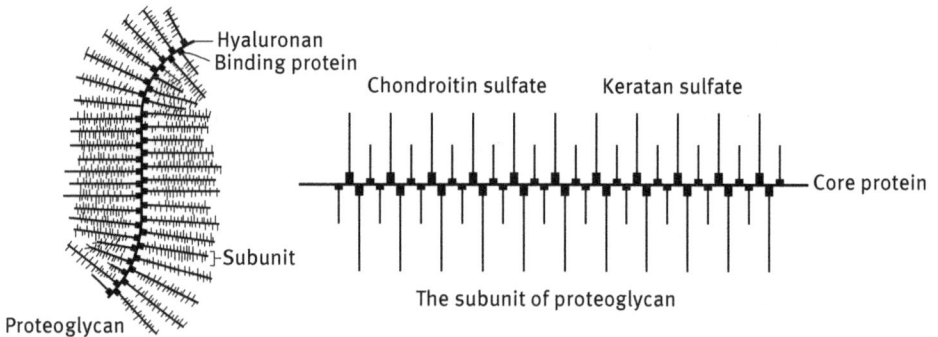

Fig. 3.1: The structure of proteoglycan.

3.1.2 Histochemical classification of carbohydrates

Histochemistry of carbohydrates (excluding immunohistochemistry and lectin histochemistry) is the histochemical technology that indirectly displays the macromolecular polymer such as glycogens, glycoproteins and proteoglycans by demonstrating some reactive groups of carbohydrate in the tissue sections. The reactive groups that can be displayed by histochemical technology mainly include glycol groups, carboxyl and sulfate groups. The glycol groups can be displayed by Periodic acid-Schiff (PAS) reaction, and the cationic dyes are used to show the carboxyl and sulfate groups.

Although carbohydrates are chemically classified into polysaccharide, glycoprotein and proteoglycan, these types of molecules are not accurately distinguished in the routine histochemical technology. According to the histochemical description and specific enzyme digestion results, the carbohydrate complexes can be divided into glycogens and glycoconjugates, including mucoitins, glycoproteins and mucoproteins. Glycoconjugates are divided into two types: acidic and neutral. The acidic glycoconjugates mainly include sulfated glycoconjugates and carboxylated glycoconjugates, and the neutral glycoconjugates originating from the epithelial tissue contain acid radicals but do not exhibit acidity by the histochemical technology.

3.2 Histochemistry methods

Many sugars are not easy to preserve because of their aqueous solubility, but most of the carbohydrates are covalently bound to proteins in tissues, and the effect of fixation occurs on the protein portions of the molecules. In most cases, aqueous or alcoholic fixative is adequate for preservation. Although not combined with proteins, the glycogens will also be well fixed because of different kinds of proteins that surround the glycogens.

The commonly used fixative includes Bouin solution, Carnoy solution, 10% formalin and 80% ethanol in the histochemical technology of carbohydrates.

3.2.1 Technology for the demonstration of glycogen

3.2.1.1 Periodic acid Schiff (PAS) reaction
The PAS reaction can be used to demonstrate carbohydrates with a number of glycol groups, which exist in glycogens (can be hydrolyzed by amylase), neutral glycoproteins (cannot be hydrolyzed by amylase) and mucins. The PAS technology is a typical reaction in histochemistry. It was created about 150 years ago, and nowadays it is still a widely used histochemical technology for the demonstration of carbohydrates.

3.2.1.1.1 Theory
Periodic acid is a strong oxidizing substance; it oxidizes glycol groups into dialdehyde groups. The PAS reaction is based on the reactivity of dialdehyde groups formed in carbohydrates with the Schiff reagent to form bright, red and insoluble compounds.

3.2.1.1.2 Methods

Preparation of the solution

a. 1% Periodic acid solution

Periodic acid	1 g
Deionized or distilled water	100 mL

b. Schiff reagent

Basic fuchsin	1 g
Double distilled water (DDW)	100 mL
Potassium (sodium) metabisulfite	2 g
1 mol/L hydrochloric acid (HCl)	20 mL
Active carbon powder	500 mg

Procedures: Dissolve 1 g of basic fuchsin in 100 mL boiling distilled water. Sufficiently mix the solution and cool it to 50°C. Add 2 g of potassium (sodium) metabisulfite and 20 mL of 1 mol/L hydrochloric acid, and keep it in a dark place at room temperature. Add 500 mg of active carbon powder and shake for 1 minute to absorb the impurities in basic fuchsin. Filter the solution and collect in a bottle. The filtered solution should be clear and chartreuse. Store the solution in the dark at 4°C and use within half year. Discard when the solution turns pink.

c. Solution of hydrosulfite

10% sodium (potassium) metabisulfite	5 mL
1 mol/L hydrochloric acid	5 mL
DDW	90 mL

Tissue processing

The tissue is fixed by formalin liquid or Carnoy solution and then paraffin embedded and sectioned.

Procedures

a. The sections are dewaxed in xylene and treated with graded ethanol to deionized water.
b. Oxidize the sections with 0.5%–1% periodic acid solution for 2 to 5 minutes.
c. Thoroughly rinse the sections with distilled water.
d. Incubate the sections with Schiff reagent for 15 minutes.
e. Rinse the sections in bisulfite solution for 2 minutes, three changes.
f. Rinse the sections in running water for 5 minutes.
g. Rinse the sections in DDW for 1 minute.
h. Stain the nuclei with hematoxylin to improve contrast.
i. The sections are dehydrated in graded ethanol and cleared with xylene and covered by coverslip with resinous mounting medium.

3.2.1.1.3 Control experiment

PAS reaction blocked by acetylation
The sections in control groups are immersed in the mixed solution containing 16 mL acetic anhydride and 24 mL pyridine for 1 to 24 hours at 22°C.

Diastase hydrolyze
The sections are hydrolyzed with 1% diastase hydrolyze solution for 40 minutes at 37°C (or for 60 minutes at room temperature). Saliva can replace the diastase: saliva is diluted 5 to 10 times in distilled water and then used to treat the sections for 30 minutes, three times. (A drop of 1% glacial acetic acid on the tip of the tongue will accelerate the salivary secretion.)

3.2.1.1.4 Results
PAS-positive substances (glycogen, neutral mucopolysaccharide and part of the acid mucopolysaccharide) are in purplish red (Fig. 3.2), glycoprotein in reddish and mast cell in red. PAS staining is negative in control tissues.

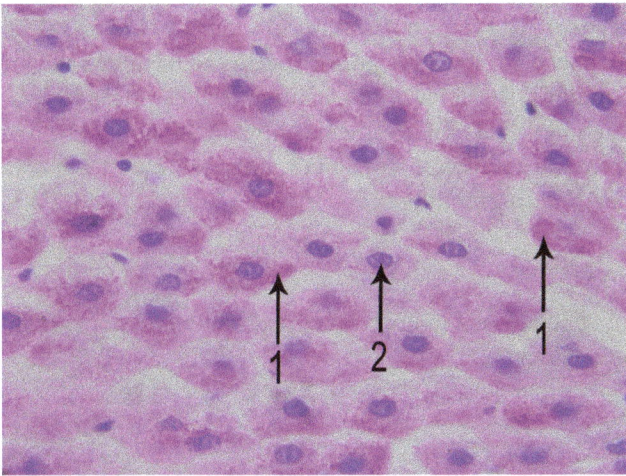

Fig. 3.2: Glycogens are PAS positive in liver cells. Arrow 1 shows the glycogen; arrow 2 shows the nucleus (hematoxylin counterstain).

3.2.1.1.5 Notes
a. The intensity of PAS reaction results is dependent to some extent on the treatment time, the pH value and the concentration of the periodic acid and Schiff reagent. The temperature increases the oxidation of glycol groups of uronic acid. If the oxidation time is more than 15 minutes, the nonspecific reaction may occur.

b. Possible free aldehyde groups within the tissue sections are likely to produce fake positives. In this case, the positive control experiments should be taken. Directly treat the adjacent section with Schiff reagent but without periodic acid oxidation, and if red color appears, it is the fake positive. In this case, sodium borohydride treatment needs to be taken to block the free aldehyde groups.

3.2.1.2 Best's (Best F. Histochemist, 1906) method
The Best's method is one of the oldest histochemical methods to demonstrate the glycogens. After staining, the glycogen granules are clear, bright and not easy to fade.

3.2.1.2.1 Theory
The active ingredient used in the Best's method is carminic acid. Its hydrogen ions in combination with hydroxyl of glycogen produce the hydrogen bonded complexes and display red color. Ammonia is solvent of carmine and can improve the pH of dye solution.

3.2.1.2.2 Reagent preparation

Best's stock solution
Carmine (2 g), potassium carbonate (1 g) and potassium chloride (5 g) are dissolved in 60 mL distilled water. The above reagents are added inside the flask and slowly heated to boiling. After 3–5 minutes, the color of the solution changes stronger gradually. Then cool the solution and add 20 mL fresh concentrated ammonia (28% or density of 0.88). Filter and store the solution in brown grinding mouth bottle at 4°C. This stock solution can be used within 2 months.

Carmine working solution
Best's stock solution (12 mL), concentrated ammonia (18 mL) and methanol (18 mL) are mixed and filtered just before use.

Best's separation solution
Methanol (20 mL), dehydrated alcohol (40 mL) and distilled water (50 mL) are mixed together.

3.2.1.2.3 Procedures
a. Dewax the paraffin-embedded tissue sections with xylene and rehydrate through graded ethanol to water.
b. The sample sections are stained by hematoxylin working solution for 10 minutes. Rinse with running water.

c. The sample sections are stained in carmine working solution for 20–40 minutes.
d. The sample sections are directly differentiated in two changes of the Best's separation solution for 50–60 seconds.
e. The sample sections are dehydrated in graded ethanol and cleared in xylene, and covered by coverslip with mounting medium.

3.2.1.2.4 Result

Glycogen particles are stained in red, and nuclei are stained blue. The neutral mucopolysaccharides and collagens show the weak pink color.

3.2.1.2.5 Notes

a. The treatment of differentiation with 1% hydrochloric acid alcohol when dyeing for nuclei is necessary because it decreases background staining and enhances glycogen staining.
b. Carmine dye should be pure, ammonia concentration should be enough when prepared and dye solution should be freshly prepared. Container used in dyeing should be sealed with a lid. After carmine staining, the section needs to be differentiated with separation solution immediately.
c. Rinse time should be as short as possible to avoid the loss of glycogens.

3.3 Display method of glycoconjugates (mucins)

Most glycoconjugate molecules have acid groups with anion, which can be combined with cationic dyes and displayed. The frequently used dye is Alcian blue.

  The sulfated and carboxylated acidic glycoconjugates have hexosamine molecules, and its acid groups bond to cationic dye to form polymer and change the absorption spectra. It is purple red when stained by blue cationic dye, a phenomenon called metachromasia, and the commonly used metachromatic dye is toluidine blue.

3.3.1 Alcian blue staining acidic glycoconjugates

3.3.1.1 Theory

Alcian blue is a cationic dye with positive charges. At pH 2.5, its cations combine with sulfate radical or carboxyl of acidic glycoconjugates to show different colors, but they do not combine with phosphate groups of nucleic acid. Given the proper solution pH, the Alcian blue is highly selective for the tissue substances and forms stable and insoluble complexes.

3.3.1.2 Solution preparation

3.3.1.2.1 Alcian blue dye (pH 2.5)

Alcian blue 8 GX	1 g
Glacial acetic acid	3 mL
Distilled water	97 mL
Thymol crystals	one grain

3.3.1.2.2 Counterstain solution of nuclei

Usually the 1% neutral red or nuclear fast red is introduced.

3.3.1.3 Procedures

a. The paraffin-embedded tissue sections are dewaxed in xylene and rehydrated through graded ethanol to deionized water. These sections are rinsed in distilled water for 1 minute.
b. The tissue sections are stained in the Alcian blue solution for 30 minutes or longer.
c. The tissue sections are rinsed in running water for 2 minutes and rinsed in distilled water.
d. The tissue sections are counterstained in nuclear fast red for 10 minutes.
e. The tissue sections are dehydrated in graded ethanol, cleared in xylene and conserved regularly.

3.3.1.4 Results

Acidulated carboxylic acid glycoconjugates and acidulated sulfuric acid glycoconjugates appear blue, and nuclei are stained red (Fig. 3.3).

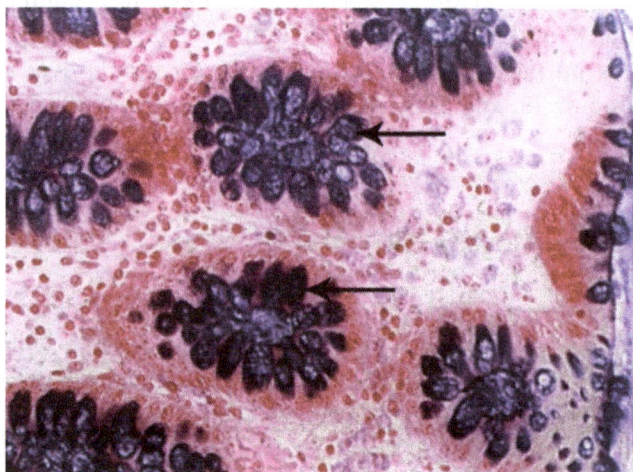

Fig. 3.3: Mucoitin (marked by arrows) in the goblet cells of large intestine stained by Alcian blue.

3.3.2 Mast cells with toluidine blue stain

3.3.2.1 Theory
Granules of mast cells contain heparin, histamine, eosinophil chemotactic factor and other chemical composition. As one of the major components, heparin belongs to sulfuric acid mucopolysaccharides, which contain acid radical and can be stained with metachromatic dye.

3.3.2.2 Reagent preparation

3.3.2.2.1 Toluidine blue stock solution
Toluidine blue O   1 g
70% alcohol   100 mL

Mix to dissolve.

3.3.2.2.2 Sodium chloride (1%)
Sodium chloride   0.5 g
Distilled water   50 mL

Mix to dissolve (make this solution fresh each time). Adjust pH value to 2.0–2.5 with glacial acetic acid or HCl.

3.3.2.2.3 Toluidine blue working solution (pH 2.0–2.5)
Toluidine blue stock solution   5 mL
1% sodium chloride, pH 2.3   45 mL

Mix well. The pH should be around 2.3 and must be less than 2.5.
    Prepare this solution freshly and discard the residue after use. pH value higher than 2.5 makes staining less contrast.

3.2.2.3 Procedures
a. The tissue samples are fixed by alcohol and acetone (1:1) for 30 minutes.
b. The tissue samples are stained by toluidine blue working solution for 2–3 minutes.
c. The tissue samples are rinsed in distilled water, three times.
d. The tissue samples are dehydrated quickly through 95% and two changes of 100% alcohol (10 seconds each because the color fades quickly in alcohol).
e. The tissue samples are cleared in xylene or xylene substitute, two times, 3 minutes each.
f. The tissue samples are covered by coverslip with resinous mounting medium.

3.2.2.4 Result

Mast cells are stained purple red. Background is stained blue (Fig. 3.4).

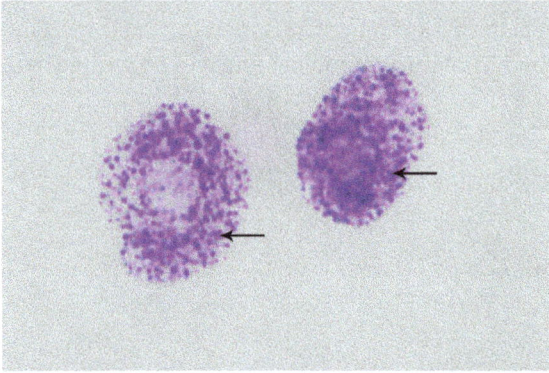

Fig. 3.4: The metachromatic granules in mast cell cytoplasm stained by toluidine blue.

Knowledge links: How does starch indicate iodine?

When starch is mixed with iodine in water, an intensely colored starch/iodine complex is formed. Many of the details of the reaction are still unknown. However, it seems that the iodine (in the form of I^{5-} ions) that gets stuck in the coils of beta amylose molecules is a soluble starch. The starch forces the iodine atoms into a linear arrangement in the central groove of the amylose coil. There is some transfer of charge between the starch and the iodine, and this changes the way in which the electrons are confined, thus changing the spacing of the energy levels. The iodine/starch complex has energy level spacing that are just so for absorbing visible light, giving the complex its intense blue color.

The complex is very useful for indicating iodine redox because the color change is very sharp. It can also be used as a general redox indicator: when there is excess oxidizing agent, the complex is blue; when there is excess reducing agent, the I^{5-} breaks up into iodine and the color disappears.

4 Nucleic acid histochemistry

Nucleic acids are the most important biological macromolecules and are essential for all known forms of life. The basic component of biological nucleic acids is the nucleotides, each of which contains a pentose sugar (ribose or deoxyribose), a phosphate group and a nucleobase (purine base or pyrimidine base). If the sugar is deoxyribose, the polymer is deoxyribonucleic acid (DNA). If the sugar is ribose, the polymer is ribonucleic acid (RNA).

The different components of nucleic acids can be demonstrated through different methods. For example, the pentose sugar can be demonstrated by Feulgen reaction, the phosphate group can be demonstrated by basic stain and the nucleobase can be determined by ultraviolet absorption spectrum. In addition, the existence of nucleic acids can also be proved through demonstrating the presence of histone.

4.1 DNA histochemical demonstration

4.1.1 Feulgen reaction

4.1.1.1 Theory
Feulgen stain is a staining technique discovered by Robert Feulgen, and it is used in histology to identify chromosomal material or DNA in cell specimens. It depends on the acid hydrolysis of DNA. Under a certain condition, the glycosidic bonds between purine bases and deoxyribose sugars are hydrolyzed and converted to aldehyde groups, which then react with the colorless reagent, the Schiff, to produce a purple compound.

4.1.1.2 Materials and reagents
a. HCl (1 mol/L)
    Hydrochloric acid   8.5 mL
    Distilled water   91.5 mL
b. Sulfurous acid solution
    10% potassium (or sodium) sulfite   5 mL
    1 mol/L HCl   5 mL
    DDW   90 mL
c. Schiff reagent (see the details in PAS reaction)

DOI 10.1515/9783110531398-004

4.1.1.3 Staining procedures

a. The tissue is fixed by 10% neutral formalin or Carnoy, paraffin embedded and sectioned.
b. Dewax the sections by passing through two changes of xylene. Rehydrate the sections by passing through 100%, 95%, 70% and 50% alcohols. Finally, rinse gently in distilled water for 2 minutes.
c. Hydrolyze the sections by 1 mol/L HCl at 60°C for 10 minutes.
d. Immerse the sections in 1 mol/L HCl at room temperature (1–2 minutes), then rinse gently in distilled water (1–2 minutes).
e. Stain the sections in Schiff reagent for 30–60 minutes (protection from light).
f. Bleach the sections with sulfurous acid (2 minutes) for three times, and rinse gently in fresh distilled water (5 minutes).
g. Dehydrate the sections with ethanol, clear with xylene and mount in resinous medium.

4.1.1.4 Results

DNA in purple-red color.

4.1.1.5 Notes

In the Feulgen experiment procedure, the hydrolysis of purine from DNA exposes the C_1 aldehydes of deoxyribose. On further hydrolyzation, the nucleic acid would be converted to histone and nucleotide; thus, the duration of hydrolysis is critical and should be strictly controlled.

The pH value of Schiff reagent should be controlled between 3.0 and 4.3.

4.1.2 The improved Feulgen reaction

4.1.2.1 Procedure

a. The tissue is fixed by alcohol formalin acetic fixative, paraffin embedded and sectioned.
b. Dewax the sections by two changes of xylene. Rehydrate the sections by 100%, 95%, 70% and 50% alcohols. Finally, rinse gently in distilled water (2 minutes).
c. Hydrolyze the sections by 1 mol/L HCl below 20°C for 40 minutes.
d. Stain the sections in Schiff reagent for 30–60 minutes (protection from light).
e. Bleach the sections with sulfurous acid (2 minutes) for three times, and rinse gently in fresh distilled water (5 minutes).
f. Dehydrate the sections with ethanol, clear with xylene and mount in a resinous medium.

4.1.2.2 Results
DNA should be stained in a purple-red color (Fig 4.1). More DNA would be preserved under lower temperature (below 20°C).

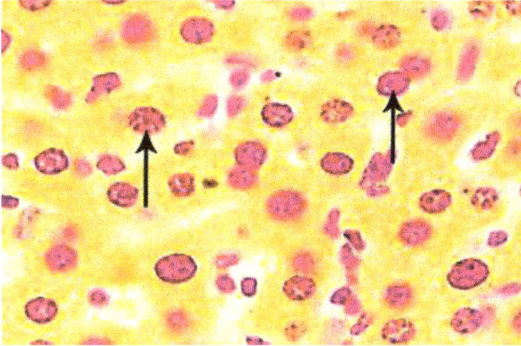

Fig. 4.1: The mouse liver cells stained by the Fulgen method show purple nuclei (marked by arrows).

4.2 Comparison demonstration of DNA and RNA

4.2.1 Methyl green-pyronin stain

4.2.1.1 Theory
Methyl green and pyronin are both water-soluble cationic basic dyes that can bind to anionic phosphate group and are mainly applied to detect DNA and RNA of cells, respectively, so that DNA and RNA are more easily viewed. DNA will be stained in green, and RNA will be stained in red. Based on this theory, we can digest RNA with ribonuclease, and the nonstaining part represents the absence of RNA, which gives the evidence to the contrary of the existence of RNA.

4.2.1.2 Materials and reagents

4.2.1.2.1 Acetate buffer (0.1 mol/L)
Solution A (0.1 mol/L acetic acid): dilute 0.59 mL glacial acetic acid to 100 mL with distilled water.

Solution B (0.1 mol/L sodium acetate): dissolve 1.36 g sodium acetate with distilled water, and then dilute it to 100 mL with distilled water.

0.1 mol/L acetate buffer: mix 80 mL solution A and 20 mL solution B. The final pH value is approximately 4.1.

4.2.1.2.2 Methyl green-pyronin stain solution
Mix the following chemicals thoroughly.

Methyl green	0.5 g
Acetate buffer	100 mL
Pyronin	0.2 g

4.2.1.3 Procedure
a. The tissue is fixed by Carnoy fixative at 4°C, paraffin embedded and sectioned.
b. Dewax the sections by passing through two changes of xylene. Rehydrate the sections by passing through 100%, 95%, 70% and 50% alcohols. Finally, rinse gently in distilled water (2 minutes).
c. Stain the sections in methyl green-pyronin stain solution at 37°C for 1 hour.
d. Dry the sections with filter paper, then acetone separation and dehydrate sections for 5 seconds (must be quick).
e. Dehydrate and clear the sections with 1:1 mixed acetone and xylene for 1–2 minutes.
f. Dry the sections with filter paper, then clear with xylene twice (3–5 minutes each).
g. Mount the sections in a resinous medium. (This progress must be quick or the color of nuclei would turns into black easily.)

4.2.1.4 Results
DNA should be stained in green color and RNA in red color. Digest RNA with ribonuclease, and the nonstaining part represents the absence of RNA, which gives the evidence to the contrary of the existence of RNA.

4.2.2 Acridine orange stain

4.2.2.1 Theory
Acridine orange (AO) molecules could enter nucleic acid bases in alive cells and form ionic bonds with phosphate radical and then show different colors for DNA and RNA.

4.2.2.2 Materials and reagents
a. Acetic acid (1%)
b. Phosphate-buffered saline (pH 6.0)
c. AO staining solution: dissolve AO (50 mg) in 40 mL distilled water, and then adjust pH to 6.0 with phosphate-buffered saline. The final fluid volume is 50 mL, and the final concentration is 0.1%.
d. Calcium chloride solution: dissolve calcium chloride (11 g) in 50 mL distilled water.

4.2.2.3 Procedure

a. The tissue is fixed by Carnoy fixative (or acetone ethanol solution at 4°C, or 80% ethanol), embedded by paraffin and sectioned (or frozen sectioned).
b. Dewax the sections by passing through two changes of xylene. Rehydrate the sections by passing through 100%, 95%, 70% and 50% alcohols. Finally, rinse gently in distilled water (2 minutes).
c. Rinse the sections in 1% acetic acid for 15–20 seconds.
d. Rinse the sections in distilled water for 10–15 seconds.
e. Stain the sections in AO staining solution for 10–15 seconds.
f. Rinse the sections in phosphate-buffered saline (pH 6.0) for 1 minute.
g. Immerse the sections in calcium chloride solution for 20–25 seconds.
h. Rinse the sections in phosphate-buffered saline (pH 6.0) again for 10–15 seconds.
i. Mount the sections in a water-soluble medium.
j. Observe and photograph under a fluorescence microscope immediately.

4.2.2.4 Results

DNA should be stained in a bright green color and RNA in red color. After nucleic acid extract or digest, the tissue is negative.

4.2.2.5 Announcements

a. Besides RNA, polysaccharides could also be stained red.
b. The tissue cannot be fixed by formalin or Bouin fixative.
c. The fluorescence from AO is sensitive and could be used in quantitative histochemistry.

Knowledge links: The study history of DNA

Although the discovery of DNA was in the 1860s, it was in the 1950s that the structures and functions of DNA were deeply studies. In the study history of DNA, the first person who noticed DNA was a 24-year-old Swiss doctor named Friedrich Miescher. In the 1868, he went to the University of Tübingen in Germany to pursue his career and began to study the chemical compositions of pus collected from hospitals. He found the sediment contained rich elements of phosphorus, which was something different from the chemicals in the previous experiments, and in the experiment, it is soluble in basophilic solution but will precipitate again when the solution turns acid. Actually, the sediment mainly contains a large molecule of DNA. Until in 1944, three scientists, Oswald T. Avery, Colin MacLeod and Maclyn McCarty, studied *Pneumovirus* and published an epoch-making research paper, which suggested that the DNA, not the protein, is the carrier of genetic information.

5 Lipid histochemistry

Lipids are one of the most important large molecules for life. The main biological functions of lipids include energy storing, signaling and acting as main molecular structures of membranous components of the cells. Phospholipids are the main structural component of biological membranes, along with lipoproteins and cholesterols. The synthesis of prostaglandin requires the unsaturated fatty acids. Steroids have different biological roles as hormones and signaling molecules.

5.1 Overview of the lipid histochemistry

5.1.1 Categories of lipids

The classification and categories of lipids had not been standardized until Bayliss-High brought out a particular way to classify the categories of lipids in 1982. The simplified classification is as follows:

5.1.1.1 Nonbinding lipids
These mainly include free fatty acids and cholesterol.

5.1.1.2 Binding lipids
a. Esters   The esters mainly include cholesterol esters and triglycerides.
b. Phospholipids   Phospholipids mainly include glycerophospholipids, cephalin and plasmologen.
c. Sphingosine base   The sphingosine base mainly include sphingomyelins, cerebrosides and gangliosides.

Neutral lipids mentioned in histochemistry mainly usually refer to triglycerides, cholesterol, cholesterol esters, steroids and some lycolipids. Acidic lipids mainly refer to fatty acids and phospholipids.

5.1.2 Fixation

Frozen sections or carbowax-embedded sections without fixation could obtain the best demonstration effect of lipids. However, in the application of histochemistry, fixation is necessary for the preservation of the fine structure of tissues. Alcohol and other lipid-soluble chemicals cannot be applied in the fixation of lipids as the lipids could be easily dissolved by alcohol. The best fixative that commonly used for lipids is Baker's calcium formalin (FCa) because phospholipids would be better preserved in the existence of calcium.

DOI 10.1515/9783110531398-005

5.2 Demonstration of lipids with physical methods

Many dyes could dissolve in lipids of the tissues, such as Sudan dyes and Oil Red O. These dyes could be used for staining neutral triglycerides and lipids on frozen sections and some lipoproteins on paraffin sections. However, these dyes are not soluble in water, but they are slightly soluble in alcohol. Thus, these dyes need alcohol as medium when applied to lipids staining. In this case, 50%–70% saturated solution is commonly used, and 60% isopropyl alcohol or 60% triethyl phospholipids is the best choice as medium for lipids staining. However, not all of the lipids can be stained by these dyes, for example, the free fatty acid and the glycerophosphoric acid.

5.2.1 Neutral lipids demonstration by Sudan black stain

5.2.1.1 Theory
Lysochromes, such Sudan black and Oil Red O, are soluble in the lipids of the tissues (fat-soluble dye), which can be applied to neutral lipids demonstration. Sudan black is formed by coupling of diazotized 4-phenylazonaphthalenamine-1 with 2,3-dihydro-2,2-dimethyl-1H-perimidine. It has the appearance of a dark brown to black powder with maximum absorption at 596–605 nm. The lipids are stained blue-black.

5.2.1.2 Materials and reagents

5.2.1.2.1 Sudan black solution
Add 0.7 g Sudan black into 100 mL propylene glycol gradually and stir thoroughly. Heat the solution to boil and maintain for several minutes, and then filter the solution with Whatman no. 2 filter paper. Filter the solution again with glass filter after cooling to room temperature.

5.2.1.2.2 Kaiser's glycerol mounting medium
Gelatin   10 mL
Distilled water 52.5 mL
Glycerol   62.5 mL
Phenol   1.25 mL

5.2.1.3 Procedures
a. The tissue should be fixed by 10% neutral formalin or Bouin then paraffin embedded and sectioned. Cryostat sections are also feasible.
b. Dehydrate the sections with pure propylene glycol for 10–15 minutes.
c. Stain the sections in Sudan black solution for 10 min.

d. Differentiate the sections by passing through 85% propylene glycol.
e. Rinse the sections gently in fresh distilled water (1–2 minutes).
f. Stain the sections in nuclear fast red for counterstaining cell nuclei.
g. Rinse the sections thoroughly in fresh distilled water several times.
h. Mount the sections in a Kaiser's glycerol mounting medium.

5.2.1.4 Results
Lipids show blue-black color, whereas cell nuclei are red.

5.2.2 Neutral lipids demonstration by Oil Red O stain

5.2.2.1 Materials and reagents
a. Stock solution – Oil Red O isopropyl alcohol saturated solution (99%)
b. Working solution

| Stock solution | 6 portion |
| Distilled water | 4 portion |

Filter the solution with Whatman no. 2 filter paper 10 minutes after mixing thoroughly. This working solution is stable for 1–2 hours.
c. Hematoxylin – as used in regular histology staining
d. Lithium carbonate (0.05%)

5.2.2.2 Procedures
a. Rinse the sections in distilled water.
b. Stain the sections in Oil Red O working solution for 6–15 minutes.
c. Rinse the sections in 60% isopropyl alcohol to clean the background as needed.
d. Rinse the sections thoroughly in fresh distilled water.
e. Stain the sections in hematoxylin for counterstaining cell nuclei.
f. Rinse the sections in fresh distilled water.
g. Immerse the sections in 0.05% lithium carbonate.
h. Rinse the sections thoroughly in fresh distilled water.
i. Mount the sections in a water-soluble medium.

5.2.2.3 Results
Lipids show red, whereas cell nuclei are blue.

5.3 Demonstration of lipids with chemical methods

Osmium tetroxide is a widely used staining agent for lipids demonstration in light and electron microscopy.

5.3.1 Theory

Osmium tetroxide (OsO_4) is one of the earliest used lysochromes. It dissolves in fats and can be reduced by organic materials to elemental osmium, an easily observed black substance. It aggressively oxidizes many materials, leaving behind a deposit of nonvolatile osmium in a lower oxidation state.

5.3.2 Materials and reagents

Osmium tetroxide solution (1%)
Osmium tetroxide	1 g
Distilled water	100 mL

5.3.3 Procedure

a. The tissue should be fixed by 10% neutral formalin, and 10–15 μm thick frozen sections are recommended.
b. Rinse the sections in fresh distilled water.
c. Stain the sections in 1% osmium tetroxide solution for 24 hours (keep in dark place).
d. Rinse the sections thoroughly in fresh distilled water.
e. Immerse the sections into anhydrous alcohol for several hours.
f. Rinse the sections thoroughly in fresh distilled water.
g. Mount in a Kaiser's glycerol mounting medium or dehydrate sections with ethanol, clear with xylene and mount in a resinous medium.

5.3.4 Results

Lipids stain black, whereas the background shows up gray to brown color (Fig. 5.1).

5.3.5 Notes

a. Osmium tetroxide is highly poisonous, even at low exposure levels, and must be handled with appropriate precautions. In particular, inhalation at low concentrations probably leads to pulmonary edema and subsequent death. Osmium tetroxide also damages the human cornea, which can lead to blindness if proper safety precautions are not observed.

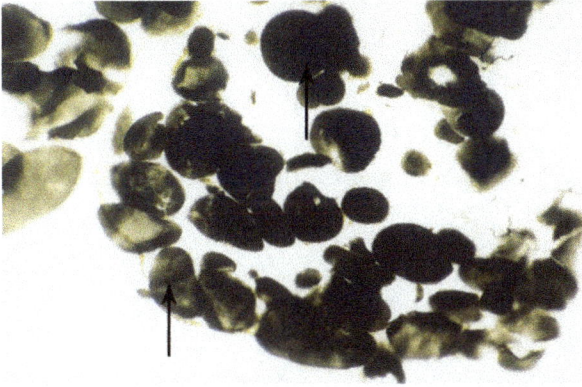

Fig. 5.1: Adipose cells (marked by arrows) stained by osmium tetroxide.

b. Osmium tetroxide is also a fixative and widely used in transmission electron microscopy to provide contrast to the image. As a lipid dye, it is also useful in scanning electron microscopy as an alternative coating.

Review question

How many kinds of lipids can be found in the human tissue, and how are they classified in biochemistry?

6 Enzyme histochemistry

Enzymes are the special catalytic proteins in organism. They are closely related to various functional activity of the human body, and the biochemical reactions such as synthesis and decomposition are also related to enzyme types and activities. Enzyme histochemistry is a science that studies the intracellular localization and the activities of enzymes in biological tissue cells by using different kinds of histochemical methods. It also focuses on illuminating the relationships between enzymes and cell structures and functions. This chapter mainly introduces the commonly used enzyme histochemical technology and their related theories.

6.1 Enzyme and its basic histochemical theory

6.1.1 Classification of biological enzymes in body

More than 2200 enzymes of various types and of different structures and functions had been found in vivo. Among them, only approximately 200 enzymes can be detected by histochemical methods. According to the standards of the International Enzyme Society and based on enzyme catalytic reaction properties, the enzymes can be divided into six categories: oxidoreductase, transferase, hydrolase, lyase, isomerase and ligase or synthetase. Among these enzymes, the histochemistry methods for oxidoreductase and hydrolase are most widely used.

6.1.2 Significance of enzyme histochemistry

Certain enzymes found in certain cells and organelles have certain particular activities. Therefore, in histochemistry, some enzymes were usually used as characteristic enzymes of certain tissues, cells and organelles. Even in the same cell, the activity of the same enzyme varies because of its functional states, such as in different development stages, in different metabolic states or pathology states, etc., so the strength of enzyme activity has also been used as one of the symbols to judge cell functions. Because of this, in enzyme histochemical technology, maintaining the exact locations and ensuring the biological activities of enzymes in samples are very necessary.

6.1.3 Histochemical reaction of enzyme

Enzyme histochemical reaction is a kind of method in which a series of reactions are involved. In the reaction process, enzymes in tissue cells act on their substrates with

DOI 10.1515/9783110531398-006

the production of reaction products under certain conditions, and then the captured agents combine with reaction products, forming colored sediments in light microscope (LM) or high electron density products in electron microscope (EM), the visible properties of which will indirectly verify the location and activities of certain enzyme (Fig. 6.1). The specific substrates and the optimal reaction conditions are both the important factors to obtain ideal results. The commonly used enzyme histochemical reactions are listed as follows.

$$\text{Substrate} \xrightarrow{\text{Enzyme}} \text{Primary reaction products} \xrightarrow{\text{+Caputure}} \text{The final reaction products}$$

Fig. 6.1: Theory of enzyme histochemical reaction.

6.1.3.1 Metal precipitation method

Metal precipitation methods are designed based on metal properties, that is, some metals itself show special colors (e.g. silver is black), or some metal compounds possess the properties of showing different colors (e.g. CoS is black, and PbS is dark brown). In these methods, enzymes in tissue cells catalyze substrates and produce the decomposition products, which will then be captured by the specific metal ions to form precipitation, and then, by using the chromogenic reaction (some need replacement reaction before chromogenic reaction), the positions of the metallic compounds will be displayed (i.e., the presences of enzymes). The basic equation is shown in Fig. 6.2.

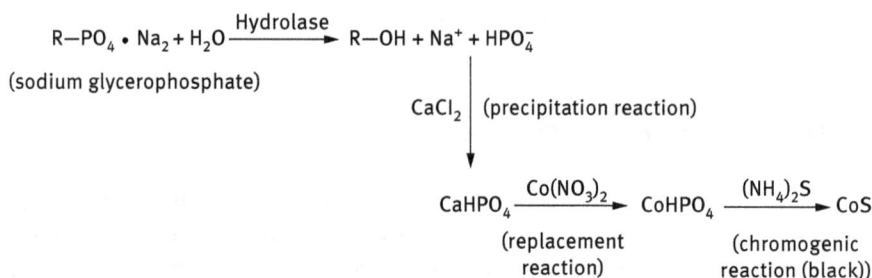

$$R-PO_4 \cdot Na_2 + H_2O \xrightarrow{\text{Hydrolase}} R-OH + Na^+ + HPO_4^-$$

(sodium glycerophosphate)

$CaCl_2$ │ (precipitation reaction)

$$CaHPO_4 \xrightarrow{Co(NO_3)_2} CoHPO_4 \xrightarrow{(NH_4)_2S} CoS$$

(replacement reaction)　　(chromogenic reaction (black))

Fig. 6.2: Theory of precipitation reaction.

Metal precipitation methods are usually used to display the hydrolases, such as alkaline phosphatase (AKP), acid phosphatase (ACP), glucose-6-phosphatase, adenosine triphosphatase, adenylate cyclase and thiamine focal phosphatase (TPPase).

6.1.3.2 Stabilized diazonium method

The basic process of stabilized diazonium method uses the artificially synthesized substrates to react with the enzymes within tissue cells first, and then the decomposition products are combined with diazonium salt, causing the diazonium nitrogen coupling reactions and generates insoluble diazonium pigments, which will finally reveal the positions of the enzymes. The basic equation is shown in Fig. 6.3.

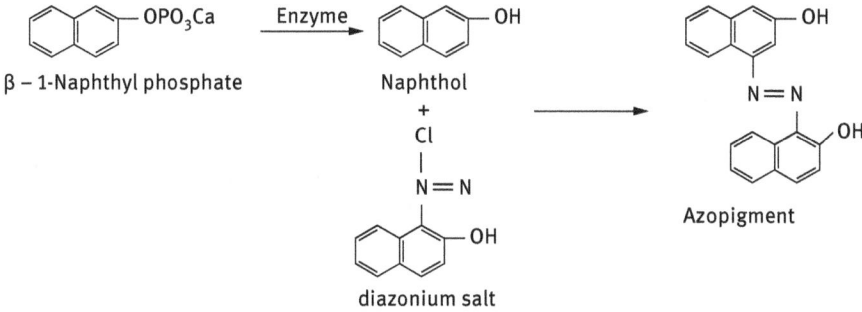

Fig. 6.3: Theory of stabilized diazonium methods.

The commonly used substrates include naphthol and naphthol AS derivatives, such as naphthol AS-BI, naphthol AS-D and naphthol AS-TR, which are often used to show phosphatase, glycosidase and lipase. Other substrates such as naphthylamine and indolamine derivatives are often used to display peptide enzyme. This method is also used for some transferase histochemistry such as γ-glutamine transferase.

Different kinds of diazonium salts (coupling agent) display different diazonium pigments colors. In enzyme histological technology, 6-azofuchsin, luxol fast blue B and permanent violet are often used. In choosing the diazonium salts, the important theories are more stabilities, faster coupling speed, lower enzyme inhibitions and smaller azopigment particles.

6.1.3.3 Tetrazolium method

Tetrazolium methods are mainly used to display oxidoreductase (dehydrogenase), such as succinate dehydrogenase (SDH), hydroxysteroid dehydrogenase, lactate dehydrogenase (LDH) and glucose-6-phosphate dehydrogenase. The basic theory is as follows: dehydrogenases in tissue cells act on specific substrates, and the hydrogen atoms are then isolated; the hydrogen atoms are transmitted by hydrogen carrier, such as coenzyme, flavoprotein or phenazine methyl ester sulfate (PMS); and finally hydrogen combined with colorless tetrazolium salt or dual-tetrazolium salt forms blue or purplish blue diformazan. The basic equation is shown in Fig. 6.4.

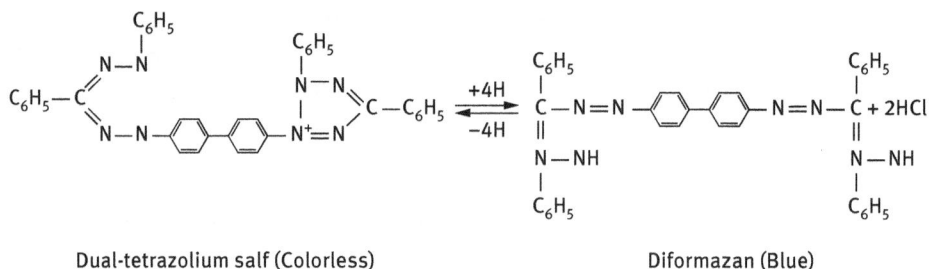

Fig. 6.4: Theory of tetrazolium methods.

Tetrazolium salts have many types. The commonly used are nitro blue tetrazolium (Nitro-BT or NBT) and tetranitro blue tetrazolium. They are light-yellow powder, soluble in water and have no inhibition effects on enzyme. They are small molecules to easily penetrate, and they have light stability, are not substantive for proteins and therefore have better histochemical effects.

6.1.3.4 Substrate film method

In substrate film method, the substrates are first dissolved into thin film and then incubated with tissue sections that contain enzymes. After that, specific staining is performed on substrate film; the unstained part (substrate has been decomposed by enzyme) is the place where the enzymes exist. This method is mainly used for the detections of amylase, hyaluronidase, sperm acrosome enzyme, etc.

6.1.3.5 Immunohistochemical method

As special protein, the antigen, the enzyme can be detected by its specific antibody, so it can be displayed by means of immunohistochemical method. The common methods are fluorescent antibody method, enzyme-labeled antibody method and immunogold sliver method. See details in Chapters 7 and 8.

6.1.4 Influencing factors of displaying effects in enzyme histochemistry method

Two main basic principles must be followed in enzyme histochemistry. One is to preserve the enzyme activity and accurate positioning of the enzymes to the largest extent in tissue cells; another is to ensure the complete tissue structure as far as possible. Therefore, the following factors should have attached importance.

6.1.4.1 Material processing

Obtaining fresh materials is the important requirement to guarantee the enzyme activity because most of the enzymes, especially dehydrogenation enzymes, are of poor tolerance to the fixative and the embedding process. Therefore, the samples in enzyme histological technology are often quick frozen by liquid nitrogen, frozen sectioned in cryostat and cold acetone fixed, or quick frozen by liquid nitrogen, fixed as thick sections at low temperature and sectioned in cryostat. Some enzyme activities have biological rhythms, such as daily rhythms, season rhythms and reproductive cycle rhythms. These factors should also be considered in the process of sampling.

6.1.4.2 Temperature
The optimal temperature should be definitely based on displaying the optimum enzyme activity. For most of the enzymes, the optimum incubation temperature range from 25°C to 37°C. In the tissues with high enzyme activities, to obtain better enzymes, the incubation temperature can be reduced correspondingly.

6.1.4.3 pH level
Enzymatic reactions of all kinds have their own suitable pH ranges. Within this pH ranges, the enzyme activities and the reaction rates are the best.

6.1.4.4 Concentration
Enzyme reaction rates are influenced by concentrations of various substances in incubation liquid, such as enzymes and substrates, reaction products, inhibitors and activators.

6.1.4.5 Inhibitors
Substances that can reduce enzyme activities are known as inhibitors. The enzyme activities in tissue sections can be inhibited by inhibitors. The inhibitors have three categories: nonspecific inhibitors, such as heat, acid and some fixatives, which have equally inhibitory effects on all enzymes; specific inhibitors, such as E600 inhibiting B esterase, tetraisopropyl phosphoramide (ISO-OMPA) inhibiting cholinesterase, etc.; and competitive inhibitor, such as propylene acid sodium inhibiting SDH, etc. For this reason, in enzyme histochemical technology, specific inhibitors or competitive inhibitor were often used in negative control group.

6.1.4.6 Activators
Substances that can enhance the enzyme activities are called activator, and the commonly used activators are Mg^{2+} and Ca^{2+}.

6.2 Histochemistry for common enzymes

6.2.1 Alkaline phosphatase (AKP)

AKP is a hydrolytic enzyme. It hydrolysis various phosphate esters of alcohol and phenol (such as sodium β-glycerophosphate and phosphoric acid naphthalene ester) in alkaline condition (pH 9.0–9.6). The hydrolytic activity could be activated by Mg^{2+}, Mn^{2+}, Zn^{2+} and Co^{2+} and easily inhibited by cysteine, cyanide and arsenic acid salt. In most tissues, the AKP could be selectively inhibited by tetramisole.

6.2.1.1 Theories
a.   Metal precipitation method (Ca-Co method, Fig. 6.5)

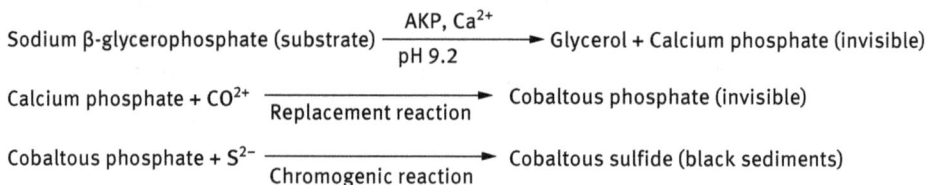

$$\text{Sodium β-glycerophosphate (substrate)} \xrightarrow[\text{pH 9.2}]{\text{AKP, Ca}^{2+}} \text{Glycerol + Calcium phosphate (invisible)}$$

$$\text{Calcium phosphate + CO}^{2+} \xrightarrow[\text{Replacement reaction}]{} \text{Cobaltous phosphate (invisible)}$$

$$\text{Cobaltous phosphate + S}^{2-} \xrightarrow[\text{Chromogenic reaction}]{} \text{Cobaltous sulfide (black sediments)}$$

Fig. 6.5: Theory of metal precipitation method to show AKP.

b.   Stabilized diazonium method (Fig. 6.6)

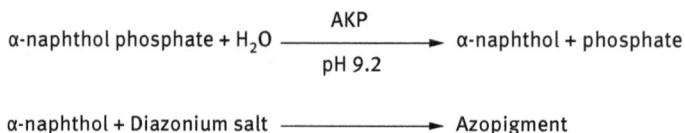

$$\text{α-naphthol phosphate + H}_2\text{O} \xrightarrow[\text{pH 9.2}]{\text{AKP}} \text{α-naphthol + phosphate}$$

$$\text{α-naphthol + Diazonium salt} \xrightarrow{} \text{Azopigment}$$

Fig. 6.6: Theory of stabilized diazonium method to show AKP.

6.2.1.2 Displaying methods

6.2.1.2.1 Ca-Co method showing AKP
Solution preparation
a.   Incubation buffer

2% Sodium β-glycerophosphate	6 mL
2% Barbital sodium	6 mL
2% Ca $(NO_3)_2$	3 mL
Distilled water	15 mL

b.   Cobalt nitrate solution (1%–2%)
c.   Ammonium sulfide (1%–2%)

6.2.1.2.2 Procedures
a.   Fresh tissue is fixed with 10% neutral formaldehyde for 12–24 hours or with 95% acetone-liquid alcohol for 4–12 hours (change the fixative for two times, room temperature)

b.   Dehydrate the sections with pure alcohol, clear with xylene, embed with paraffin and then section.
c.   Dewax and rinse the sections with water for three times.
d.   Incubate the sections at 37°C for 30 minutes to 2 hours.
e.   Rinse the sections with water and then react with 1%–2% cobalt nitrate for 2–5 minutes.
f.   Rinse the sections with water for three times, then react with 1%–2% sulfur amine for 1–2 minutes.
g.   Routine water rinse, dehydrate, clear, mount and seal the sections with resinous medium.

6.2.1.2.3 Result
The AKP active sites are dark brown or black stained. The nuclei are negative.

6.2.1.2.4 Notes
a.   Control experiments should be set in Ca-Co method. Remove the substrate from incubation buffer, or process the tissue section with inhibitor Lugol iodine solution for 3 minutes or levamisole.
b.   Paraffin-embedded tissue section can display most hydrolases. The postfixed or cryostat sections, fixed with cold acetone or 10% formalin, will get better effects; however, a shorter incubation time is needed.

6.2.1.3 Stabilized diazonium methods showing AKP

6.2.1.3.1 Solution preparation

Hexazotized pararosaniline solution
Liquid A: 4% pararosaniline solution – add hydrochloride pararosaniline 0.2 g in 2 mol/L HCl 5 mL (stored in refrigerator after fully dissolved).
    Liquid B: 4% sodium nitrite solution – add 0.2 g sodium nitrite in 5 mL distilled water.
    Mix liquid A and liquid B before using (liquid A is slowly dropped into liquid B, shaking while dropping).

Incubation buffer

Naphthol AS-BI sodium phosphate	10–25 mg
Dissolve in dimethyl formamide or dimethyl sulfoxide (DMSO)	0.5 mL
Tris-HCl (pH 8.2–9.2)	50 mL
Hexazotized pararosaniline solution	0.5 mL

Adjust the pH level to 9.2 with NaOH, fully mix and filtrate as reserve solution.

6.2.1.3.2 Procedures

a. The methods of sampling, fixation and sectioning are the same as that of Ca-Co method, steps 1–3.
b. Incubate the sections for 5–60 minutes at 37°C (for liver tissue, 60 minutes is recommended).
c. Rinse the sections with water.
d. Stain the sections with 1% methyl green counterstain for 5–10 minutes.
e. Routine mount and seal.

6.2.1.3.3 Results

Positive cytoplasms are uniformly red, negative nuclei present bluish-green (results of counterstain) and control group is negative.

6.2.1.3.4 Notes

a. In the process of preparing incubation solution, the occurrence of milk-like suspension or the colored precipitations is due to the coupling of free naphthol and diazonium salt. Filtration would help to clear it.
b. The colors of azopigments depend on the types of diazonium salt; for instance, solid blue B, solid blue BB, RR, VRT and VB are bluish violet, and solid red TR, solid purple B, hexazotized pararosaniline solution and hexazotized magenta III are red.

6.2.2 Acid phosphatase (ACP)

ACP belongs to the hydrolytic enzyme, which widely exists in body tissue. The optimum pH level of its hydrolysis reactions ranges from 5.4 to 6.0. Inhibitors of enzymes show organs and species differences. The prostate ACP is inhibited by tartaric acid and fluorides, but not by 0.5% formalin; the liver ACP can be inhibited by the three inhibitors previously mentioned; and the red blood cell ACP can be inhibited by formalin and fluorides, but not by tartaric acid. However, in most of the tissues, ACPs can be selectively inhibited by NaF.

6.2.2.1 Theory

6.2.2.1.1 Metal precipitation methods
For lead methods, see Fig. 6.7.

Sodium β-glycerophosphate + H_2O $\xrightarrow[\text{pH 4.98}]{\text{ACP } Pb^{2+}}$ Glycerol + Lead orthophosphate (invisible)

Lead orthophosphate + S^{2-} ⟶ Lead sulfide (brownish black sediments)

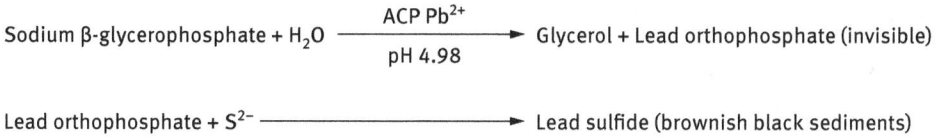

Fig. 6.7: Theory of metal precipitation method to show ACP.

6.2.2.1.2 Stabilized diazonium methods
The principles are the same as that of stabilized diazonium methods of AKP.

6.2.2.2 Lead methods showing ACP

6.2.2.2.1 Preparation of solutions
a. Acetate barbiturate buffer
    Sodium acetate   9.714 g
    Sodium barbital   14.714 g
    Distilled water   500 mL
b. HCl (1 mol/L)
c. Incubation buffer
    Acetic acid barbiturate buffer   1.7 mL
    8.5% NaCl   0.7 mL
    0.1 mol/L HCl   3.6 mL
    Distilled water   24 mL
    3.3% Lead nitrate   0.25 mL
    3.2% Sodium β-glycerophosphate   2.6 mL (drops while shaking)
    0.6 mol/L $MgSO_4$ (activator)   0.7 mL

Adjust pH level (0.1 mol/L HCl) to 4.7–4.8.

6.2.2.2.2 Staining procedures
a. The preparation of section is as the same as that of AKP.
b. The sections are dewaxed and immersed in water.
c. Incubate the sections at 37°C for 2–6 hours. Rinse the sections with distilled water three times, 2 minutes each.
d. Immerse the sections into 1%–2% ammonium sulfide for 1–2 minutes. Rinse the sections with distilled water three times, 2 minutes each.
e. Routine dehydration, clearance and sealing.

6.2.2.2.3 Results

ACP-positive reactions are black (Fig. 6.8). Control group (omit the substrate or add inhibitors 0.01 mol/L NaF) is negative.

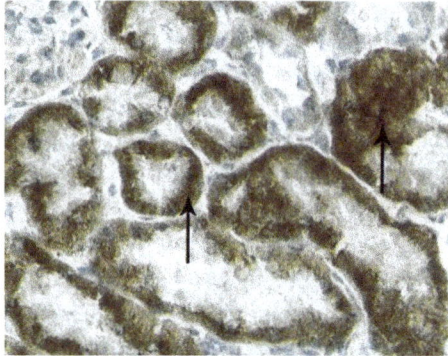

Fig. 6.8: Lead phosphatase method shows ACP activities (marked by arrows) in rat renal tubules.

6.2.2.2.4 Notice

a.   ACP is a kind of soluble enzyme; cold fixation is preferred.
b.   Sediments should not present in liquid preparation process.
c.   The pH level of incubation buffer must be accurate, no more than 5.2.
d.   Lead ions have inhibitory effects to the ACP, and nonenzyme proteins have non-specific adsorption to the lead ions, so the incubation time should be shortened as much as possible.

6.2.3 Succinate dehydrogenase (SDH)

SDH belongs to the oxidoreductase category. Dehydrogenases are divided into two groups, among which group 1 do not need coenzyme dehydrogenase (such as SDH and glycerol-3-phosphate dehydrogenase), whereas group 2 needs coenzyme dehydrogenase (such as hydroxyl steroid dehydrogenase, glucose-6-phosphate dehydrogenase, LDH and malate dehydrogenase).

6.2.3.1 Theories (Fig. 6.9)

Lactic acid	NAD⁺	PMSH	Methyl thiazolyl tetrazolium
Acetonate	NADH	PMS	Difornazam (Blue sediments)

Fig. 6.9: Theory of NBT method to show SDH.

The optimum pH level is 7.6; the competitive inhibitor is malonic acid sodium; in addition, –SH reagents such as mercury, selenium, parachloromercuribenzoic acid (0.01–0.1 mmol/L) and fluorides are also inhibitors of SDH. SDH is sensitive to fixative, so fresh tissue sections should be chosen or fixed as well in cold formaldehyde for 5–15 minutes.

6.2.3.2 Tetrazolium salt method showing SDH

6.2.3.2.1 Preparations of incubation buffer

0.2 mol/L phosphate buffer (pH 7.6)	12 mL
Nitro-BT (4 mg/mL)	2.5 mL
0.2 mol/L sodium succinate	12 mL

6.2.3.2.2 Staining procedures for fresh tissue
a. Fresh tissue is cryostat sectioned.
b. Keep the sections for 15–60 minutes at room temperature or 4°C, 1–2 hours (the time depended on the reaction conditions).
c. Rinse the sections with normal saline for 10 minutes, fix with formaldehyde for 10 minutes and rinse again with water.
d. Immerse the sections in 80% alcohol for 5 minutes.
e. Seal the sections with glycerin gelatin.

6.2.3.2.3 Staining procedures for frozen tissue
a. Fresh tissues are frozen in liquid nitrogen for 1–2 minutes.
b. Cut the samples into thick sections in –15°C in the cryostat (100–300 microns).
c. Fix the sections for 20 minutes with 1%–2% paraformaldehyde or 0.25% glutaraldehyde (4°C).
d. Rinse the sections for 60 minutes with phosphate buffer (4°C) and change after three rinses.
e. Incubate the sections for 1–3 hours at 37°C.
f. Rinse the sections with saline for 30 minutes.
g. Postfix the sections with 10% FCa for 30 minutes and then rinse.
h. The sections are dehydrated with graded alcohol, cleared and embedded.
i. Routine preservation.

6.2.3.2.4 Control experiment
a. Omit the substrate.
b. Add malonic acid sodium (final concentration is 3.7 mg/mL) in incubation buffer.

6.2.3.2.5 Results
The places where the enzyme is active show purple blue; the control group is negative.

6.2.3.2.6 Notice
a. In dehydrogenase histochemistry, DMSO is regarded to have functions of stabilizing the dehydrogenase and cellular structure, increasing substrates penetration and helping the electron transfer, therefore enhancing the staining effect.
b. The respiratory inhibitors (100 mmol/L cyanide or azide) in incubation buffer can prevent the oxygen in the atmosphere from competitive combining to H in reaction buffer, and also have the effects of improving the histochemical reaction results.

6.2.4 Lactate dehydrogenase (LDH)

LDH is an indispensable coenzyme dehydrogenase. With the presence of coenzyme I (NAD), LDH participates in the process of lactic acid and pyruvic acid interconversion. It is soluble and widely distributed and plays a key role in a variety of cell metabolism processes. Because of its two molecule polypeptide chains, M and H, LDH forms five isozymes: M4, M3H1, M2H2, M1H3 and H4. They vary in distribution in different tissues.

6.2.4.1 Theories (Fig. 6.10)

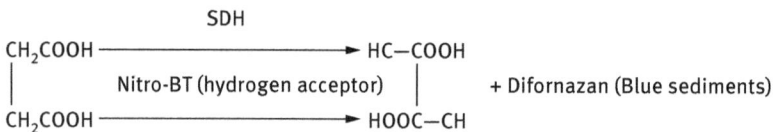

```
                     SDH
CH₂COOH ─────────────────────────► HC─COOH
  |          Nitro-BT (hydrogen acceptor)  |      + Difornazan (Blue sediments)
CH₂COOH ─────────────────────────► HOOC─CH
```

Fig. 6.10: Theory of NBT method to show LDH.

6.2.4.2 Tetrazolium salt method showing LDH

6.2.4.2.1 Preparations of incubation buffer

1 mol/L D,L-sodium lactate	0.1 mL
NAD (4 mg/mL)	0.1 mL
0.06 mol/L phosphate buffer (pH 7.0)	0.25 mL
NBT (4 mg/mL)	0.25 mL
PMS or 1-methoxy PMS	1.96 mg
0.1 mol/L KCN	0.1 mL
0.5 mol/L MgCl₂	0.1 mL
Distilled water	10 mL

6.2.4.2.2 Staining procedures
a. Fix the tissue in 2% precooled paraformaldehyde, rinse with PBS and section in vibration microtome or cryostat.
b. Incubate the sections 5 minutes in the dark at room temperature for 30 minutes.
c. Rinse the sections with normal saline, dehydrate, seal with glycerin gelatin and then examine under microscope.
d. Control experiment: omit the substrates from incubation buffer.

6.2.4.2.3 Results
The positive enzyme reaction positions are stained blue (Fig. 6.11), and the control test is negative.

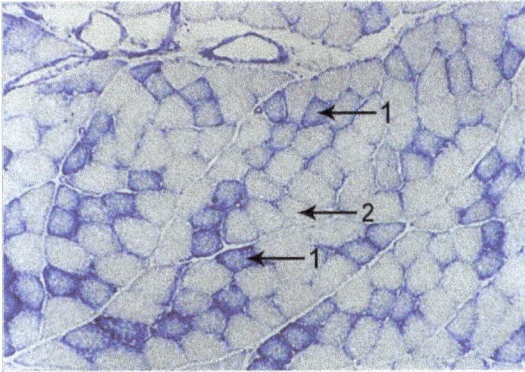

Fig. 6.11: Tetrazolium salt method shows LDH activities in skeletal muscle cells (arrow 1, positive; arrow 2, negative).

6.2.4.2.4 Notice
LDH is well tolerated to aldehyde fixative, so the fixed tissue enzymes are not easy to spread and can be accurately positioned.

Review question

How should alkaline phosphatase, glucose-6-phosphatase, pyrophosphate thiamine enzymes, adenosine triphosphatase, nonspecific esterase, 3β-hydroxyl steroid dehydrogenase and cytochrome oxidase be demonstrated?

7 Basic theory of immunohistochemistry

Immunohistochemistry is an important branch of histochemistry. It is a technique to identify the cellular or tissue constituents by means of specific antigen-antibody reactions, and the site of antibody binding is visualized by visible labeling of the antibody. At present, the term immunohistochemistry and immunocytochemistry are often used interchangeably.

7.1 Basic immunology

Specific antigen-antibody reactions are the immunohistochemical crucial links.

7.1.1 Antigen

An antigen is the substance that can stimulate the body to produce specific immune response.

7.1.1.1 Basic properties of the antigen

Antigen has two basic characteristics: one is immunogenicity that means the antigens can stimulate the body to generate antibodies and sensitized lymphocytes and, thus, induce a humoral and/or cell-mediated immune response; the other is immunoreactivity, which means the antigens can combine specifically with antibodies and sensitized lymphocytes. The complete antigens, such as most of the proteins and individual macromolecular polysaccharides, have both characteristics of immunogenicity and immunoreactivity. The substances that have the immunoreactivity but the lack of immunogenicity are known as incomplete antigen or haptens, which include the vast majority of polysaccharides, lipids, small molecular peptides and some simple chemical substances.

7.1.1.2 Character of antigen

7.1.1.2.1 Foreignness

In general, antigen is a heterogenic or allogeneic substance. The more xenogeneic the foreign molecules, the stronger the immunogenicity. The similar proteins and polypeptide antigen that belonged to same categories share same or similar epitope in common; hence, cross-reactions may occur. For example, the antiserum of rabbit antihuman LHβ can react with pituitary LH of a variety of mammals.

DOI 10.1515/9783110531398-007

The immune system normally discriminates between self and nonself components; thus, foreign molecules are immunogenic. However, in particular cases, for example, self tissue changes and becomes "nonself" substance (as degenerative cells, virus-infected cells and cancer cells). Alternatively, because the body's barrier is damaged, exposing the isolated components to the immune system (such as sperm, eye crystal protein and thyroglobulin), these tissue components may become autoantigens and are attacked by the immune system, leading to autoimmunity.

7.1.1.2.2 Physicochemical property

Antigens are macromolecules. Usually, their molecular weight are greater than 10 kDa. The higher the molecular weight of the substance, the stronger the immunogenicity. The immunogenicity of spherical molecular protein is better than the immunogenicity of fibrous proteins. The proteins in aggregation state have stronger immunogenicity than that of monomeric proteins.

To obtain immunogenicity, hapten needs to attach to macromolecules such as proteins. This protein is called carrier. In the hapten-carrier conjugates, carrier determines immunogenicity and hapten decides the antigenic specificity of the conjugates. However, the carrier may also produce antibodies against itself.

7.1.1.2.3 Specificity

The specificity refers to the ability of antigen to react with corresponding immune response substances (antibody or sensitized lymphocyte) specifically. The basis of specificity is the special chemical groups of surface of the antigens and their spatial configuration, also known as antigenic determinant or epitopes. In general, the antigenic determinants are limited to approximately three to eight amino acid residues. In polysaccharide molecules, there are approximately five to seven monosaccharide residues, and in nucleic acid molecule, there are approximately six to eight nucleotide residues. Antigenic determinant depends not only on the primary structure of the antigen molecule but also on the three-dimensional configuration. Antigenic determinant hidden within the antigen molecules can be unmasked by enzyme digestion to increase them to react with the antibodies.

7.1.2 Antibody

Antibody is also called immunoglobulin (Ig). There are five types of Ig found in the human body: IgA, IgD, IgE, IgG and IgM. IgG is the most abundant in the serum. From the functional perspective, the antibody binds specifically to the corresponding antigen. IgG can activate complements and selectively pass through the placenta barrier to reach and protect the fetus.

7.1.2.1 Basic structure and properties of antibody

7.1.2.1.1 Basic structure

IgG has a four-chain structure; it is composed of two identical long polypeptide chains (440 amino acid residues, also known as heavy chains, H chains) and two identical short polypeptide chains (214 amino acid residues, also known as light chains, L chains) linked by disulfide bonds to form symmetrical a Y-shaped structure. In the carboxyl terminal of the polypeptide chain (C-terminus), three-quarters of the heavy chain and one-half of the light chain form a constant region (C-region) where the amino acid sequence is constant in the same species. In the amino terminal of the polypeptide chain (N-terminus), one-quarter of the heavy chain and one-half of the light chain form variable region (V region), where the amino acid sequence highly varies. This is the basis of the diversity of antibodies and the body to adapt to a variety of antigen stimulation to produce diverse specific antibodies. The antibody combines with the specific antigen located in the V region. Conversely, it is also the V region that makes the antibody specificity.

IgG, IgD and IgE own the basic units mentioned previously, but the difference is the amino acids of heavy chain. IgM is a pentamer with five basic units. IgA is generally a dimer containing two basic units and is usually secreted with exocrine secretions.

7.1.2.1.2 Antigenicity

Antibody is macromolecular substance, and its molecular weight is approximately 150 kDa. According to the antigenicity of heavy chain, Ig is divided into five types. The antigenicity of heavy chain is denoted by Greek letters: γ, δ, α, μ and ε. They determine the type of Ig: IgG, IgD, IgA, IgM and IgE, respectively. IgG, IgM and IgA can be further divided into different subclasses. IgG can be divided into four subcategories (IgG1, IgG2, IgG3 and IgG4), IgM and IgA are divided, respectively, into two subcategories (IgM1 and IgM2, and IgA1 and IgA2). The structure of Ig light chain is similar and has two types: lambda (λ) and kappa (κ). Each antibody contains two light chains that are always identical, and only one type of light chain, κ or λ, such as the structure of IgG may be $\gamma_2\kappa_2$ or $\gamma_2\lambda_2$. Because κ and λ of all antibodies are the same, that is, they have a common antigenicity, there will be a cross-reaction on the serology.

7.1.2.1.3 Functional fragments

Antibody molecules consist of two parts: antigen binding fragments (Fab) and crystallizable fragments (Fc) (Fig. 7.1). The narrow area connecting Fab and Fc is called the hinge region. Each Fab is composed of one complete light chain and one incomplete heavy chain. The variable domain of Fab contains the sites that can combine specifically with antigens. Thus, one antibody could combine with two antigens. The Fc is composed of two incomplete heavy chains. It plays a role in modulating antibody biological activity, such as IgG binding complements and crossing the placenta barrier and IgE binding receptors on mast cell membrane. Fc mainly embodies the antigenicity of antibody because the amino acid sequences in Fc are very conservative in evolution. All the antigenicity of antibodies in all the individuals of the same species is the same and can be detected by its antibody. For example, different kinds of antigens are selected to immunize the rabbits to produce different kinds of rabbit-derived antibodies (known as primary antibodies), and the rabbit-derived antibodies are collected from serum, purified by biochemistry methods and injected into a goat as antigens to generate goat-derived antibodies (known as secondary antibodies). In this way, the secondary antibodies will specifically combine with all kinds of primary antibodies whose Fc regions are the same.

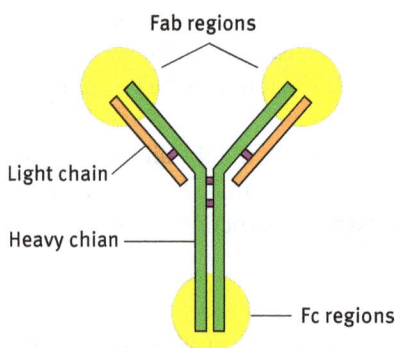

Fig. 7.1: Functional fragments of antibody.

7.1.2.2 Production of antibodies
According to different antibody preparation methods, the antibodies used in the immunohistochemical techniques are divided into polyclonal antibody and monoclonal antibody.

7.1.2.2.1 Polyclonal antibodies
When antigens are used to immunize a mammal, multiple antigenic determinants of antigen stimulate the multiple plasma cells in body to produce various antibodies, and these mixed antibodies are known as polyclonal antibody. Polyclonal antibodies with similar antigen (containing some of the same or similar antigen determinants) may cause cross-reaction. Because the polyclonal antibody separate from animal serum, they are also called antiserum.

7.1.2.2.2 Monoclonal antibody
When antigens are used to immunize a mammal, multiple antigenic determinants of antigen stimulate the multiple plasma cells in body to produce various

antibodies. The multiple plasma cells are isolated from the animal, and one plasma cell will fuse with a myeloma cell, which does not produce antibodies but can continually proliferate. The fused cells are called hybridomas. These hybridomas will continually grow and secrete antibodies in culture. Single individual hybridoma cell is isolated and cultured and keep secreting the antibodies that are against one kind of determinants of antigen, and these antibodies are called monoclonal antibodies.

7.1.2.3 Antigen-antibody reaction
When the corresponding antigen and antibody encounter, the specific binding reaction occurs, like the relationship between the lock and the key. High specificity, high affinity and insoluble products are the basis of immunohistochemistry. Dissociation coefficient of antigen-antibody complex is generally 10^{-11} mol/L.

7.2 Common markers and their detection

Marker is the precondition for an immunohistochemical reaction product to become visible. In immunohistochemical technique, common markers include fluorescent dyes, enzyme, biotin, metal, etc. They are mainly attached to an antibody but can also be labeled to antigen, protein A, lectins, etc.

7.2.1 Fluorescent dyes

7.2.1.1 Common fluorescent dyes
Fluorescent dye (also known as fluorochrome) is a substance that can emit character fluorescence under stimulation from a high-energy light (ultraviolet light). Now the fluorescent dyes used in labeling antibodies include fluorescein isothiocyanate (FITC), tetramethyl rhodamine isothiocyanate (TRITC), Texas red, phycoerythrin R, carbocyanine (Cy) as Cy2, Cy3, Cy5, etc. FITC is yellow-orange or isabelline crystalline powder and is soluble in water and ethanol solvent. Its molecular weight is 389.4, its maximum absorption wavelength of light is 490–495 nm and its maximum emission wavelength is 520–530 nm. It presents bright yellow-green fluorescence. FITC is stable and can be stored at room temperature in a dry place for more than 2 years. TRITC is an amaranth powder. Its maximum absorption wavelength of light is 550 nm, its maximum emission wavelength is 620 nm and it presents orange-yellow fluorescence. The carbocyanine fluorescent dyes are widely used in recent years because of its strong fluorescence and high resistance to fluorescence quenching. Fluorescent markers can be observed directly by fluorescence microscope.

7.2.1.2 Theory of preparation fluorescence-labeled antibody

The isothiocyanate of FITC or TRITC can be linked covalently with free amino groups of antibody to form sulfur-carbon-amino bonds that conjugate antibodies with fluorochrome. The reactions occur at alkaline aqueous solution. The equation is shown in Fig. 7.2.

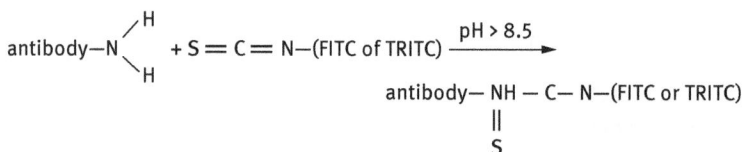

$$\text{antibody}-N\begin{smallmatrix}H\\ \\H\end{smallmatrix} + S=C=N-(\text{FITC of TRITC}) \xrightarrow{pH > 8.5}$$

$$\text{antibody}- NH - \underset{\underset{S}{\|}}{C}- N-(\text{FITC or TRITC})$$

Fig. 7.2: Preparation equation of fluorescence-labeled antibody.

One antibody can be labeled by 15–20 FITC molecules. FITC is commonly used in single staining, and TRITC is usually introduced together with FITC for double-staining techniques.

7.2.2 Enzyme

7.2.2.1 Enzyme for labeled antibody

At present, the enzymes are still frequently used markers in the immunohistochemical. Enzyme for labeling should satisfy the following conditions: the substrate must be specific and easy to display; the end products of reaction must be insoluble, stable and not easy to diffuse; the enzyme is easily obtainable in a highly purified form and stable property; and the enzyme-labeled antibody does not hinder the catalytic activity of the enzyme and the immunological activity of the antibody. Endogenous enzymes should not be present in the subject tissue; if any, endogenous enzymes of the same type must be inhibited. Horseradish peroxidase and alkaline phosphatase are now widely used as an antibody labels.

7.2.2.2 Preparing principle of enzyme-labeled antibody

Aldehyde groups of glutaraldehyde covalently bind to amino groups of enzymes and antibodies, forming enzyme-labeled antibody. The equation is shown in Fig. 7.3.

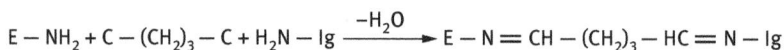

$$E - NH_2 + C - (CH_2)_3 - C + H_2N - Ig \xrightarrow{-H_2O} E - N=CH - (CH_2)_3 - HC=N - Ig$$

Fig. 7.3: Preparation equation of enzyme-labeled antibody.

7.2.2.3 Demonstrations of enzyme marker

7.2.2.3.1 Theory

Enzymes are biological catalysts, having the specificity for substrate. Provided the specific substrates, proper pH level and temperature, the demonstration of enzyme activity is often achieved by the location of the end products of enzymatic reaction. The reaction product must be insoluble, and the end products must have color or high electron density, which makes them visible under light microscope or electron microscope (Fig. 7.4).

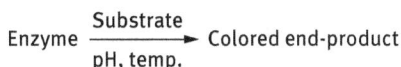

$$\text{Enzyme} \xrightarrow[\text{pH, temp.}]{\text{Substrate}} \text{Colored end-product}$$

Fig. 7.4: Demonstrations of enzyme marker.

7.2.2.3.2 Demonstration for peroxidase

When hydrogen peroxide exists as substrate, the peroxidase will produce oxygen, which oxidizes the electron donor to produce colored end products (Fig. 7.5).

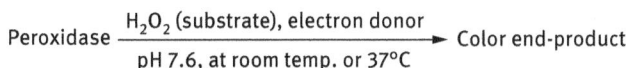

$$\text{Peroxidase} \xrightarrow[\text{pH 7.6, at room temp. or 37°C}]{H_2O_2 \text{ (substrate), electron donor}} \text{Color end-product}$$

Fig. 7.5: Demonstration of peroxidase.

At present, the frequently used electronic donors include diaminobenzidine (DAB), 4-chloro-1-naphthol (CN), 3-amino-9-ethylcarbazole (AEC) and tetramethyl benzidine (TMB), producing brown, blue, red and blue reaction end products, respectively.

DAB development process

DAB is oxidized to produce a dark brown, insoluble precipitate that will be observed under the light microscope and can be preserved permanently. The reaction end product can be made electron dense for electron microscopy by treatment with osmium tetroxide. Any other electron donors do not exhibit these advantages better; thus, although DAB has been reported to be a potential carcinogen, it is still widely used.

The DAB solution makeup method is as follows:
a. Dissolve 50 mg DAB in 100 mL 0.1 mol/L, pH 7.6 Tris buffer (TB).
b. Add 33 μL 30% H_2O_2.

Mix well, filter and add H_2O_2 Final concentration: 0.05% DAB, 0.01% H_2O_2, 1–5 minutes.

CN development process

CN is not carcinogenic; hence, it can be used as a replacement for DAB or associate with DAB for double staining. The product dissolves in organic solvents and is not resistant to dehydration and xylene treatment. Therefore, it cannot be permanently preserved, so a water-based mountant (such as glycogelatin) must be used. The photograph should be taken quickly after staining because the reaction products are easy to fade and tend to diffuse.

The CN solution makeup method is as follows:

a. Dissolve 3 mg of CN in 0.5 mL of anhydrous alcohol.
b. Add 9.5 mL of 0.1 mol/L TB, mix.
c. Add 3% H_2O_2, three drops.
d. Filter if precipitate form.
e. Develop at room temperature for 5–10 minutes. Enhance coloration if heated to 50°C.

AEC development process

This method gives a red end product. It is introduced as an alternative to DAB or useful in multiple immunostaining. Disadvantages and notes are the same as the CN.

The AEC solution makeup method is as follows:

a. Dissolve 20 mg of AEC in 0.5 mL of dimethyl formamide.
b. Add 9.5 mL of 0.05 mol/L acetate buffer, mix.
c. Add 3% H_2O_2 three drops.
d. Develop at room temperature, 5–10 minutes.

TMB development process

TMB has high lipid solubility, which helps it enter the cell easily. TMB is oxidized to form coarse, dark blue deposits, which are insoluble in organic solvents. TMB may be a potential carcinogen.

The TMB solution makeup method is as follows:

a. Acetate buffer: 190 mL of 1 mol/L HCl in 400 mL of 1 mol/L sodium acetate solution. Make up to 100 mL with double distilled water. Adjust pH level to 3.3.
b. Solution A: Dissolve 100 mg of nitro potassium ferricyanide in 5 mL of acetate buffer. Add 92.5 mL double distilled water.
c. Solution B: Add 5 mg of TMB in 2.5 mL of anhydrous alcohol, heat to dissolve.
d. Mix together solutions A and B, add 10–50 μL of 30% H_2O_2. Develop for 20 minutes.

7.2.2.3.3 Display for alkaline phosphatase

Alkaline phosphatase releases the orthophosphate and naphthol derivatives from the naphthol phosphate substrate, and naphthol derivatives combine with Fast Blue BB to form the insoluble blue azo dyes, so that the activity is specifically located. If Fast Blue BB is replaced with Fast Red, the end products are red (Fig. 7.6).

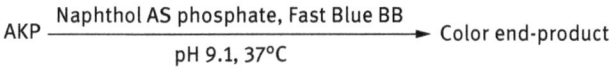

$$\text{AKP} \xrightarrow[\text{pH 9.1, 37°C}]{\text{Naphthol AS phosphate, Fast Blue BB}} \text{Color end-product}$$

Fig. 7.6: Demonstration of alkaline phosphatase.

Blue end product
a. Dissolve 2 mg of naphthol AS-MX phosphate or naphthol AS-TR phosphate in 0.2 mL of dimethyl formamide and then add 9.8 mL of 0.1 mol/L TB (pH 8.2) quickly while stirring (store at 4°C for several weeks). Before use, add Fast Blue BB to the 1 mg/mL (filter if necessary).
b. The previous solution is dropped onto the sections for 10–15 minutes to check microscopically for the blue reaction product.
c. Rinse in Tris-buffered saline (TBS) and terminate reaction.
d. Mount in glycerogelatin.

Red end product
As in the previous section, use Fast Red TR salt instead of Fast Blue BB.

Permanent red end product
a. Dissolve 10 mg naphthol AS-TR phosphate in 0.2 mL of dimethyl formamide, and then immediately add 40 mL of 0.2 mol/L TB (pH 9.0).
b. Before use, mix 250 μL of 4% new fuchsin (prepare with 2 mol/L HCl) in 250 μL of 4% sodium nitrite solution and then add in substrate solution.
c. Check pH level and then filter. Add on the sections. Incubate at room temperature for approximately 10 minutes.
d. Rinse in running water. Counterstain the nucleus if necessary.
e. Dehydrate rapidly, clear and mount permanently.

7.2.3 Biotin

7.2.3.1 Biotin and avidin
Biotin, or vitamin H, is water soluble and exists in the egg yolk, liver, etc. It is a heterocyclic, S-containing monocarboxylic acid. Its molecular structure is simple (Fig. 7.7). It can label antibody, enzyme, etc., via its carboxyl group in combination with the amino of protein.

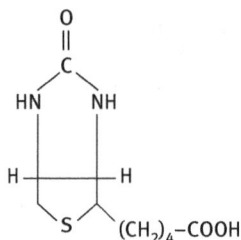

Fig. 7.7: Chemical structure of biotin.

Avidin is composed of four identical subunits. Each of them specifically binds one molecule of biotin; thus, a total of four biotin molecules can bind to a single avidin molecule. Avidin has a high affinity for the biotin (KD 10^{-15} mol/L). The bond formation between biotin and avidin is very rapid, and once formed, the conjugate is very stable, and the dissociation only occurs at pH 1.5. Avidin may be labeled with other markers, such as fluorescein, enzyme and colloidal gold. Using labeled (or unlabeled) avidin and biotin in biotin-labeled antibody (or enzymes) in high affinity binding, the corresponding specific antigen can be detected.

In addition, biotin can also be labeled by lectin, hormones and nucleic acid probe. Avidin-biotin system with high sensitivity can be used in affinity histochemical and *in situ* hybridization histochemical technique for detecting specific polysaccharides, hormone receptors and specific mRNA (or DNA).

7.2.3.2 Preparation of biotinylated antibodies
a. Preparation of antibodies: Dissolve antibodies in 0.1 mol/L sodium bicarbonate buffer to the final concentration l mg/mL, and dialyze for 24 hours.
b. Preparation of biotin: Biotin for marking must be activated (esterification).
c. Marker: Add 60 µL of activated biotin in 1 mL of antibody and mix well.
d. Dialyze for 1–2 days and centrifuge to remove the precipitate.

7.2.4 Colloidal gold

Colloidal gold is typically used for electron microscopy immunolabeling. It is an electron-dense, noncytotoxic, stable cytochemical marker. Colloidal gold has hydrophobic and negative charge on the particle surface, which maintains its stability from the electrostatic repulsion. Colloidal gold is usually either a red color (for spherical particles less than 80 nm) or a blue color (for larger, nonspherical or agglomerated particles). Proteins, which usually have positively charged groups, are attracted to the negatively charged colloidal gold particles via the electrostatic interaction to form a stable complex. In this way, the antibody, protein A and other macromolecules can be labeled with gold particles.

Because the electronic density of gold particles is very high, they can be used for transmission electron microscope without special treatment. Excited by electron beam, the gold particles emit secondary electrons; thus, they can be directly used for scanning electron microscope. Gold particles larger than 10 nm are pink, blue or purple under light microscope; thus, they can also be used for light microscopy.

7.3 Basic conditions

The basic conditions for immunohistochemistry reaction usually include the preservation of the organizational structure and antigen, specific antibody production and effective marks.

7.3.1 Storage of antigens and unmasking of antigens

Try to preserve the natural tissue structures and antigens as well as their antigenicity in the process of tissue preparation.

7.3.1.1 Fixation
The antigen to be tested must be insoluble or become insoluble by fixing, which prevents the peptide and protein antigens to diffuse.

7.3.1.1.1 Fixative
The frequently used fixatives include ethanol, acetone, formaldehyde, glutaraldehyde or their mixture. Ethanol and acetone are often used for rapid fixed tissue in cryostat sections to detect cell surface antigens and autoantibodies or Ig inside and outside the cell. Formaldehyde is the most widely used fixative; 1%–4% dissolved paraformaldehyde in phosphate buffer (PB) can be used alone or together with other fixative. It is mainly used for the detection of peptides and proteins.

Tissue fixed in glutaraldehyde will be more extensively cross-linked than that in tissue fixed in formalin. The extensive cross-linking adversely affects immunohistochemical staining but does provide excellent ultrastructural preservation. Thus, it is extensively used as a primary fixative for electron microscopy. Among the mixed fixative, the Bouin solution and the Zamboni solution are more commonly used.

7.3.1.1.2 Fixation methods
For cell surface antigen, the frozen section must be used and fixed in acetone. Monolayer cells, viruses and bacteria are commonly fixed in cold acetone or ethanol; for example, add fixative in cell suspension and concentrate them by low-speed centrifugation to prepared specimen. The soluble antigens can be fixed using steam formaldehyde.

7.3.1.2 Embedding and sectioning

Commonly used methods of specimen preparation include vibrating sectioning, frozen sectioning, paraffin sectioning and stretched preparation.

7.3.1.2.1 Vibrating sectioning

Use a vibrating razor blade to cut through tissue. Previously fixed or fresh tissue pieces are embedded in low-temperature agarose (or without using the agarose to embed). Place a small dot of superglue on the workstation, and carefully place the cut edge of the tissue block on the superglue. Wait a few minutes to fully adherence. Fill the bath with buffer until tissue block is totally immersed. Choose appropriate vibration amplitude and speed to cut into 20–100 μm thick piece. Individual sections are then collected with a fine brush or curved glass needle in the buffer. In general, vibration sections are stained by floating staining.

7.3.1.2.2 Frozen sectioning

The frozen sectioning includes cryostat sectioning, semiconductor refrigerating frozen sectioning and CO_2 refrigerating frozen sectioning. The cryostat sectioning is better. The fresh, unfixed or fixed tissue destined for cryotomy should be frozen as quickly as possible at low temperature (–70°C or less). Then the tissue is transferred into the cryostat to warm up to the cutting temperature (approximately –20°C) before sectioning. Place the tissue on top of the cryostat chuck with forceps and cut into 4–20 μm thick sections. Frozen sections are transferred to slides coated with adhesive agent. With fingers touching the back of the section, the contrast in heat helps to "melt" the frozen sections on to the slide. To prevent the formation of ice crystals, the tissue can be immersed into an appropriate sucrose solution overnight. In recent years, frozen ultrathin sectioning can be used in electric immunohistochemical technique.

7.3.1.2.3 Paraffin sectioning

Although the tissue is processed through organic solvents and high temperature, which may destroy the antigenicity, it is still adopted for most routine and special stains. In addition, sensitivity of the immunohistochemical method is greatly improved. Thus, paraffin continues to be the most popular infiltration and embedding medium in the histology laboratory.

7.3.1.2.4 Stretched preparation

Some thin layer tissue such as the iris, mesentery, small vessels and others can be directly stretched. Gastrointestinal tract, heart, etc., can be dissected and stretched layer by layer in order to observe all parts where antigen-antibody reaction occurred. The fixation will be taken when required.

7.3.1.3 Increase the contact of antigen and antibody
To improve the sensitivity of immunohistochemical staining, the chance of antigen-antibody interaction must be increased.

7.3.1.3.1 Proteolytic enzyme digestion
When formalin-based fixatives are used, intermolecular and intramolecular cross-linkages are formed with certain structural proteins; these are responsible for the masking of the tissue antigens. Antigen reactivity can be recovered by protease digestion, which is called unmasking. Commonly used enzymes are trypsin and pronase. The mechanism of unmasking is unknown.

7.3.1.3.2 Antigen retrieval
Some antigen detection needs to be improved by high temperature and high pressure for paraffin section. Its mechanisms are poorly understood. This may be related to the heat-induced cross-linking fracture, denatured proteins dissolution and other relevant factors. At present, the commonly used antigen retrieval methods are as follows:

a. Microwave antigen retrieval method: Dewax and rehydrate the paraffin section. Place the slides in the plastic staining jar containing 0.01 mol/L, citric acid buffer, pH 6.0. The buffer solution must cover the slides completely. Place them in the domestic microwave oven with medium power for approximately 10 minutes until the solution boils.
b. Pressure cooker antigen retrieval method: Add a certain amount of antigen retrieval buffer in the pressure cooker. Carefully place slide rack into the hot solution and seal the lid. When the pressure cooker reaches full pressure, keep it for 10 minutes. After cooling, rinse sections in phosphate-buffered saline (PBS).

The harsh conditions of antigen retrieval induce tissues to shed; thus, a strong adhesive agent is needed.

7.3.1.3.3 Increase the penetrability of tissue and cell
If the cell membrane is integral, macromolecules cannot pass through. The following methods can be used to increase the penetrability of tissue and cell:

a. Use the nonionic surfactant (also called the detergent), such as polyethylene glycol octylphenol ether (Triton X-100) and saponin, to dissolve the lipid.
b. Dehydrate and rehydrate the section twice.
c. Multigelation: Put the specimen treated by a sucrose solution into the cryogenic refrigerator to rapidly freeze. Remove the specimen from cryogenic refrigerator, and then place it in the laboratory until completely melted. Repeat previous procedures for two to three times.

7.3.2 Specific antibody

Primary antibody is a specific antibody against the antigen. A secondary antibody is a specific antibody that binds to primary antibody. The success of immunohistochemical staining depends on the quality of these antibodies primarily, especially the quality of primary antibodies.

7.3.2.1 Request for specific antibody

Antibodies must have high specificity, high affinity and high titer. Specificity refers to the ability of an individual antibody to react with an antigenic determinant or the ability of a population of antibody molecules to react with only one antigen. Antibodies should be considered as the detection reagent for site or region, instead of a specific reagent for antigen. Some of these antibodies may cross-react with other molecules. Cross-reactivity refers to the ability of an individual antibody to react with more than one antigenic determinant or the ability of a population of antibody molecules to react with more than one antigen. Thus, sometimes an antibody derived from one animal species, which is injected by some peptide antigens, can be used to detect the same peptide antigen from another species animal.

Antibody affinity is the strength of the reaction between a single antigenic determinant and a single combining site on the antibody. The higher the affinity of the antibody for the antigen, the more stable the interaction. The longer the epitope sequence, the higher the affinity of the corresponding antibody, the less the chance of cross-reaction and the less the opportunity of being rinsed out during the treatment. If the antibody titer in the antiserum is higher, the antiserum can be highly diluted, and the most of the interference can be disregarded.

7.3.2.2 Serum antibody purification

When negative reaction appears after the antiserum has been absorbed by the specific antigen, when the antiserum can be highly diluted or when the background staining is very slight, the antiserum can directly be applied in the reaction without further purification; however, cross-reactions should still be closely observed. If there was a satisfactory specific staining, but the nonspecific reaction is obvious and could not be removed by antigen absorption, the antiserum purification should be performed. The common methods include affinity chromatography purification and tissue dry powder absorption; the latter is often used in fluorescent antibody purification.

7.3.2.3 Diluents and storage of antibodies

7.3.2.3.1 Optimum dilution for antibody

It can only be established after the antibody is treated in a series of dilutions on positive control section. In general, the 0.05- to 0.1-mol/L, pH 7.2–7.6, PBS or TBS will be used as antibody dilution. When testing, the diluted concentration of antibody increases two times in the series of testing solution; for example, antibody dilutions of 1:50, 1:100, 1:200 to approximately 1:20,000 are used to stain the sections first, and then choose the optimum antibody dilution ration that causes the best staining effect to finish formal experiment.

7.3.2.3.2 Add proteins

Usually, a certain amount of protein is added in antibody diluents, such as 1% suitable normal serum (the same animal that generates the secondary antibody) or 0.1% bovine serum albumin (BSA). These proteins provide a high concentration of nonantibody proteins that can compete with the antibody for nonspecific binding sites in the tissue or combine with the impure antibodies that may exist in antiserum; thereby, nonspecific staining is reduced.

7.3.2.3.3 Storage of antibodies

If the unlabeled primary antibody is not diluted or lower diluted (1:10–1:50), they can be stored frozen at –20°C or less. The antibodies should be divided into suitable aliquots before storing in the freezer. This avoids repeated thawing and freezing that damage the activity of the antibody. Usually, the ABC kit is stored at 4°C

7.3.3 Reagents

7.3.3.1 Buffer

7.3.3.1.1 TB and TBS

a. Stock solution (TB, 1 mol/L, pH 7.6)

Tris	60.57 g
HCl 1 mol/L	210 mL
Double distilled water (DDW)	Make up to 500 mL

Dissolve Tris in 200 mL double distilled water, add HCl and adjust pH to 7.6 with 1 mol/L HCl or 1 mol/L NaOH. Finally, add double distilled water to 500 mL and store in the refrigerator at 4°C.

b. TB, 0.1 mol/L, pH 7.6: Take 10 mL stock solution and add 90 mL double distilled water, and adjust pH level to 7.6. This is mainly used for the preparation of color reagent in immunoenzyme technique.

c. TBS, 0.1 mol/L, pH 7.6

TB (1 mol/L)	10 mL
NaCl	0.9 g
Double distilled water	Make up to 100 mL

Dissolve NaCl in 10 mL TB stock solution, make up to 100 mL with double distilled water and adjust pH level to 7.6. This reagent is widely used in immunoenzyme histochemistry technique. It can also be used to dilute the antibody and other reagents and for rinsing specimens.

7.3.3.1.2 PB and PBS
a. PB, 0.1 mol/L, pH 7.6

First, 0.1 mol/L NaH_2PO_4 and 0.1 mol/L Na_2HPO_4 are prepared, and then the two solutions are mixed according to a certain proportion to make different pH levels (see Appendix A).

This solution is used to prepare the fixative and sucrose solution.
b. PBS, 0.1 mol/L, pH 7.4

0.1 mol/L PB	100 mL
NaCl	0.9 g

Dissolve NaCl in PB and check the pH level. This solution, as well as TBS, is widely used to rinse the sample and to dilute immune reagents. Its concentration can be 0.01–0.05 mol/L.

7.3.3.1.3 Other buffer, see Appendix A.

7.3.3.2 Fixatives
a. Paraformaldehyde solution (4%)

Paraformaldehyde	4.0 g
PB, 0.1 mol/L, pH 7.4	100 mL

Add paraformaldehyde into PB, stir and heat on magnetic heating stirrer to approximately 60°C. Continue stirring until paraformaldehyde is completely dissolved and solution becomes clear. In the end, add PB to 100 mL and filter the solution.
b. The Bouin solution and the modified Bouin solution (see Chapter 2).
c. Zamboni solution

Paraformaldehyde	2.0 g
Saturated picric acid	15 mL
0.1 mol/L, pH 7.4, PB	75 mL

Dissolve paraformaldehyde in PB, and after cooling, add the saturated picric acid.
d. Periodate-lysine-paraformaldehyde fixative

Stock solution A: Dissolve 1.82 g lysine hydrochloride in 50 mL of double distilled water. Adjust pH level to 7.4 using 0.1 mol/L Na_2HPO_4. Make up to 100 mL with 0.1 mol/L, pH 7.4 PB, and this is 0.1 mol/L lysine. Store at 4°C. This solution is stable for 2 weeks.

Stock solution B: Add 8.0 g paraformaldehyde in 100 mL of double distilled water, heat and dissolve as before. Filter and store at 4°C.

To make fresh solution on daily use, mix three parts solution A and one part solution B, then add sodium periodate (also $NaIO_4$). Final concentrations: 0.01 mol/L $NaIO_4$, 2% paraformaldehyde and 0.075 mol/L lysine.

7.3.3.3 Slide adhesion
a. Chrome alum gelatin solution

Gelatin	0.1–0.5 g
Chrome alum	0.05 g
Double distilled water	100 mL

Heat double distilled water to boil. Add gelatin and stir to dissolve completely. Cool to 40°C and add chrome alum, stir to dissolve and then filter the solution. Immerse the slides in the chrome alum gelatin solution for several minutes. Take out the slides and vertically dry at 37°C in the incubator. If sections are thicker, the concentration of gelatin should be higher.

b. Polylysine

Polylysine	10 mg
Double distilled water	100 mL

Dissolve polylysine in double distilled water to make 0.01% solution. Apply a small drop to one end of a clean slide and push it into thin film on glass slide, like making blood smear. Dry in air.

7.3.3.4 Protease digestive solution
a. Trypsin (0.1%)

Trypsin	0.1 g
Calcium chloride ($CaCl_2$)	0.1 g

Dissolve $CaCl_2$ in 100 mL double distilled water. Adjust the pH to 7.8 using 0.1 mol/L NaOH. Add trypsin and stir until dissolve completely and filter the solution. Heat the digestive solution to 37°C before using. $CaCl_2$ helps dissolve trypsin and also helps in the trypsin function.

b. Pepsin (0.4%)

Pepsin	0.4 g
0.1 mol/L HCl	100 mL

Dissolve pepsin in 0.1 mol/L HCl, and heat to 37°C before using.

c. Pronase (0.06%)

Pronase	0.06 g
0.1 mol/L TB (pH 7.6)	100 mL

7.3.3.5 Detergents

7.3.3.5.1 Triton X-100
a. Stock solution (30% Triton X-100)

Triton X-100	28.2 mL
0.1 mol/L TBS (pH 7.6) or 0.01 mol/L PBS (pH 7.4)	72.8 mL

5.1.2 0.1%–2.0% Triton X-100

The stock solution is diluted by PBS or TBS to a final concentration of 0.1%–2.0%; 1%–2% Triton X-100 can be used to rinse, 0.3% is commonly used in antibody dilution and 0.1%–0.3% is commonly used in electron microscopy technology.

7.3.3.5.2 Saponin
a. Stock solution (1% saponin)

Saponin	1 g
0.1 mol/L TBS (pH 7.6)	100 mL

Stir on a magnetic stirrer to completely dissolve and store at 4°C.

b. Saponin (0.1%)

1% Saponin	1 mL
0.1 mol/L TBS (pH 7.6)	9 mL

Mix, 0.01% used for electron microscopy specimens.

7.3.3.6 Sucrose solutions
The commonly used concentration is 5%–30%, prepared with PB or TBS.

Sucrose	5–30 g
0.1 mol/L, PB (pH 7.4)	Make up to 100 mL

The solution can store in refrigerator at 4°C for 1 month. Specimens can undergo through low to high concentrations of sucrose for several hours or overnight. Specimen that is placed in a high concentration of sucrose will often float on top at first. After they sink to the bottom, they are ready to be cut.

7.3.3.7 Mountant

7.3.3.7.1 Glycerol buffer
a. Glycerin-PBS

Glycerin	9 mL
PBS 0.01 mol/L, pH 7.2	1 mL

Mix and stand by at 4°C, and this can be used after elimination of air bubbles.

b. Glycerin-carbonate buffer
Glycerin 9 mL
Carbonate buffer 0.5 mol/L, pH 9.5 1 mL

Glycerol buffer mounting media are mainly used for immunofluorescent staining specimens. It can delay the fluorescence fading and improve fluorescence brightness.

7.3.3.7.2 Glycogelatin
Gelatin 15 g
Double distilled water 100 mL
Glycerin 80 mL
Thymol a little

Heat and dissolve gelatin completely in double distilled water. After filtration, add the glycerin and mix well; a little thymol is added as a preservative. The glycogelatin is mainly used for mounting of immunoenzyme technique reaction products but cannot be processed by alcohol, xylene or CN.

7.3.3.7.3 Permanent mountants
They are used for mounting slides after DAB development, such as DPX, cedar wood oil or Canada balsam.

Knowledge links: The biotin

It took more than 40 years to discover biotin. In 1901, Wildiers found there is a necessary chemical for yeast to grow, and he called this chemical "bios." In 1916 and in 1927, Bateman and Boas, respectively, found that the rats in their experiments developed dermatitis when fed by raw egg white but the cooked egg white did not cause dermatitis. In 1936, German scientists Kogl and Tonnis found a necessary chemical for yeast to grow in cooked egg yolk and called the chemical "biotin." In 1937, Hungarian scientist Gyorgy found a kind of chemical that can neutralize the effect of raw egg white and called the chemical "vitamin H." In 1940, Gyorgy and his research members verified that bios, biotin and vitamin H actually have the same substance, and it is a necessary nutrition for mammals. The chemical structure of biotin was discovered in 1942 and was artificially synthesized in 1943.

8 Commonly used methods in immunohistochemistry

In 1941, Coons and his colleagues labeled pneumonic diplococcus mucopolysaccharides antibody with fluorescence to check pneumonic diplococcus on the mice lung biopsy. Their work established the real foundation of immunohistochemical technology. The original technology was to use isocyanate fluorescence to directly mark the specificity antibodies (primary antibodies); thus, it is called the direct method. The following indirect method is more sensitive because the fluorescence is used to label the second antibody, and then gradually, the fluorescence isothiocyanate (FITC) is used to replace isocyanate fluorescence because FITC can make the process easier and more practical. The emergence of isothiocyanate rhodamine (RITC) makes it possible to detect two antigens (double staining) because the red fluorescence launched by RITC and the yellow-green fluorescence emitted by FITC create very obvious contrast.

Antibodies marked by the enzyme method (Nakane and Pierce, 1966) appeared because the immunofluorescence method cannot clearly show the detailed structures and the samples cannot be stored for a long time. Initially, the mainly used enzyme as marker was peroxidase, followed by alkaline phosphatase and glucose oxidase. However, the activities of both enzymes and antibodies were influenced when they combine with each other. Sternberger set up the unlabeled antibody method in 1970. This method can connect the primary antibody with antienzyme antibody with bridge antibody. Later they created the peroxidase-antiperoxidase (PAP) method in which the enzymes and the antienzyme antibodies were used, and this greatly improved the sensitivity of the immunoenzyme method. In 1981, Hsu and his colleagues introduced the avidin-biotin system into immunohistochemistry and immunoenzyme histochemistry to create the avidin-biotin-peroxidase complex (ABC) method, which is more sensitive than the PAP method, and this was based on the high affinity between avidin and biotin. Nowadays, both the PAP method and the ABC method are widely used. The colored final product of immunoenzyme technology can be observed in natural light and is also visible under electron microscope by osmium oxidation.

Under the electron microscope, the ultrastructure is usually invisible because of the immunoenzyme reaction product. In 1971, Faulk and Taylor introduced immunogold staining (IGS), which checks antigens of the cell surface with colloidal gold-labeled antibody under electron microscope. Because of the simplicity and the efficiency of colloidal gold technology, it has been gradually used to label antibody, protein A and lectin in recent years.

DOI 10.1515/9783110531398-008

8.1 Theories of different methods

Immunohistochemical methods are varied but can be summed up as direct and indirect methods. According to the basic principle of application, this technology mainly has the direct and indirect methods, the unlabeled antibody enzyme assay and labeled antigen methods, etc., according to the principles (Fig. 8.1).

8.1.1 Direct method

The direct method refers to the markers that combine with the primary antibody. It is directly added to the cell or tissue preparation to react with the corresponding antigens. This is generally used only for immunofluorescence double staining method.

8.1.2 Indirect method

The primary antibody is not marked, whereas the secondary antibody is marked by fluorescence, enzyme or colloidal gold. It has higher sensitivity. The secondary antibody (like goat against rabbit IgG) with a tag can locate various antigens, provided the antigens in the tissue react with the primary antibodies that come from the same animals, for example, the rabbit. The indirect method is widely used in immunofluorescence and IGS. In addition, protein A can combine with the Fc region of the primary antibody; thus, with fluorescence, enzyme or colloidal gold marker, protein A can replace the secondary antibody in the indirect method.

By using the indirect method, the enzyme-marked antibody is formed by covalent binding of the two; thus, it has certain influence on the activity of enzymes and antibodies. In addition, if there is nonspecific antibody in the antiserum, the enzymes may also be tagged, thus increasing the background staining.

8.1.3 Unlabeled antibody enzymatic method

The feature of unlabeled antibody enzymatic method is that all the antibodies used are unlabeled. It makes the secondary antibody as the bridge connecting the primary antibody and the final antibody, and the final antibody is the antienzyme antibody produced by animal families with same origins of the primary antibody. By far, the most commonly used unlabeled antibody enzymatic method is to connect the first antibody and the antibody marker detection reagent compounds with the bridge antibody. The latter includes peroxidase-anti perioxidase (PAP) complex, alkaline-phosphatase-antialkaline-phosphatase (APAAP) compounds and ABC compounds.

8.1.3.1 PAP method

The main characteristic of the PAP method is to make compound with peroxidase and antienzyme antibodies in advance. In the dyeing process, the first layer is the primary antibody (such as the rabbit IgG), the second layer is unlabeled secondary antibody (goat antirabbit IgG) and the third layer is PAP complex (rabbit). The second layer, or the goat against rabbit IgG, must be overdosed to make sure that it connects the primary antibody with the one Fab section, and another Fab segment can combine with PAP complex of the third layer.

8.1.3.2 APAAP method

The basic theory of APAAP method is the same as that of the PAP method, and the only difference is that alkaline phosphatase replaces peroxidase.

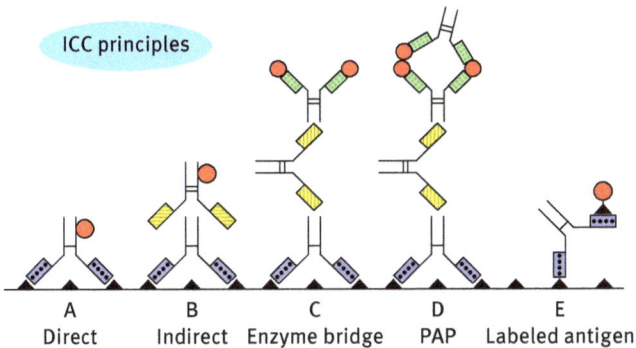

A	B	C	D	E
Direct	Indirect	Enzyme bridge	PAP	Labeled antigen

Fig. 8.1: Basic theory of immunohistochemistry.

8.1.3.3 Avidin-biotin complex (ABC) method

The ABC method is set up with similarity to the PAP method by applying the high affinity between avidin and biotin and the characteristics of biotin-labeled antibody and enzyme. The first step is to make ABC compound by mixing avidin and biotin-labeled peroxidase according to proper proportion. This results in three of the binding sites of avidin being occupied by biotin-labeled enzymes, leaving one binding site unoccupied. After the biopsy was processed by the primary antibody, it is added with biotin-labeled secondary antibody, and the primary and the secondary antibodies can combine through antigen-antibody reaction, whereas the biotin will combine with the unoccupied binding site in the ABC compound of the third layer.

8.1.3.4 Labeled antigen method

In the labeled antigen method, antigens are labeled first with enzymes, colloidal gold or radioactive isotopes. Then they are mixed with certain amount of specific antibodies to create compounds to make one Fab segment of the antibody combine with the antigen and leave another Fab unoccupied. This compound is used to detect the same antigen in tissue.

8.2 Immunofluorescence method

Immunofluorescence histochemistry was one of the most widely used and the earliest immunohistochemistry technologies. It is quick and easy, especially practical in clinical pathological diagnosis. In addition, immunofluorescence staining is often used to stain the unfixed frozen sections. Two different fluorescence can be differentiated without interference by changing filters in the double immunofluorescence staining. The application of confocal laser scanning microscopes and a new generation of fluorescent pigment (e.g., CY series) contribute greatly to the development and application of immunofluorescence histochemical technologies because they are capable of providing excellent images and vividness of the colors. Immunofluorescence, of course, also has its obvious limitations; for example, it needs special fluorescence microscopes, and it is quite difficult to show the details. In addition, fluorescent tags fade easily so the specimens cannot be preserved for a long time, and electron microscopes cannot be applied for observation.

8.2.1 Theories

First, the already-known antigens or antibodies are marked with fluorescent pigment. Then the fluorescence-labeled antibodies or antigens are used as the probe to check the corresponding antigens or antibodies in tissues or cells to form antigen-antibody compound *in situ*. Antigens or antibodies in tissues or cells can be qualitatively determined and localized by observing the fluorescence triggered by irradiation of the laser onto the fluorescence.

Immunofluorescence methods can be classified into direct, indirect and complement methods:

a. The direct method (see Section 8.1)
b. The indirect method
    i.   Antigen detection method (see Section 8.1)
    ii. Antibody detection method: In tissue or cell specimen preparation containing already-known antigen, add some need-to-know serum. The antibodies will combine with the corresponding antigens if the serum contains certain antibodies (such as autoantibodies or pathogens). Under fluorescent microscope, unique fluorescence can be observed from the spots in tissue slides where antigens locate when the antibodies combine with the antigens and react with the added fluorescence-labeled antibodies.

8.2.2 Staining methods

8.2.2.1 Direct method

a.   The section should be rinsed with 0.01 mol/L PBS, pH 7.4, for 10 minutes and then placed horizontally in wet boxes.
b.   Add fluorescent antibody at 37°C or room temperature for 30 minutes.
c.   PBS rinses, 5–10 minutes, twice.
d.   Seal the section with buffered glycerol.

8.2.2.2 Indirect method

8.2.2.2.1 Antigen detection method

a.   The section should be rinsed with 0.01 mol/L PBS, pH 7.4, for 10 minutes and then placed horizontally in the wet box.
b.   Add the primary antibody at 37°C or room temperature for 30 minutes.
c.   PBS rinses, 5–10 minutes, twice.
d.   Add the fluorescence-labeled secondary antibody at 37°C for 30 minutes.
e.   PBS rinses, 5–1 minutes, twice.
f.   Seal the section with buffered glycerol.

8.2.2.2.2 Autoantibody detection method

a.   Make complex frozen sections (4–5 μm) with liver, heart, kidney of rats, stomach of mice, human thyroid and striated muscle. The sections can be fixed with methanol or unfixed. Store the sections in –20°C. When it is time for use, take them out and rinse with PBS.
b.   Add some serum of the patient, dilute with PBS (1:5) at 37°C for 30–60 minutes.
c.   PBS rinses, 5–10 minutes, twice.
d.   Add fluorescence-labeled antihuman IgG at 37°C for 30 minutes.
e.   PBS rinses, 5–10 minutes, twice.
f.   Seal the section with buffered glycerol.

8.2.2.2.3 Complement method

a.   Rinse the section for 10 minutes with PBS.
b.   Mix equal quantity of the primary antibody (56°C, 30 minutes to inactivate complement in advance) and complement (fresh guinea pig serum 1:10 dilution in PBS, pH 7.1, containing 0.001% $MgSO_4$), add onto the section at 37°C for 30 minutes.
c.   PBS rinses, twice.
d.   Add resistance to complement the fluorescent antibody at 37°C for 30 minutes.
e.   PBS rinses, twice.
f.   Buffer glycerol seals slides.

8.2.3 Control experiments

8.2.3.1 Spontaneous fluorescence control experiments
Sometimes there is spontaneous fluorescence, which can interfere with the observation of the special fluoresce in slides, coverslips, reagents or tissue samples. Thus, the first thing to do is to make sure there is no spontaneous fluorescence. In this way, only the PBS is added to the samples then sealed with buffered glycerol. The result of fluorescence microscope observation should be negative. It verifies that PBS and buffered glycerol do not have spontaneous fluorescence.

8.2.3.2 Positive control experiment
Positive fluorescence should appear if the known samples are stained with influenza virus antigen or antibody. This can prove that both the reagents used and the methods are effective.

8.2.3.3 Alternative control experiment
Replace the primary antibody with normal serum or PBS, and the results should be negative.

8.2.3.4 Inhibition experiments
a. Two-step process: Unmarked specific antibodies are added to the samples to combine with the corresponding antigens; then add some fluorescence-labeled antibody to acquire negative reaction or significant weakening of the fluorescence because of the saturation of the binding sites on the antigens.
b. One-step process: The results should be negative if the fluorescent antibody and the unlabeled antibody are equally mixed and then add them onto the specimen. This method is simple and convenient, and the effect is comparatively good.

8.2.4 Observing and recording of the results

Immunofluorescence staining sections must be observed or photographed with fluorescent microscope. Strict instructions should be followed to operate a fluorescence microscope. Lasers with different excitation wavelengths should be used to match the different fluorescent dyes to obtain positive staining images with varied colors (Fig. 8.2). This should generally be performed in a dark room, and immediate photography should be taken after staining. To prolong the storage period by postponing the fading of the fluorescence, the specimen should be sealed and stored at 4°C.

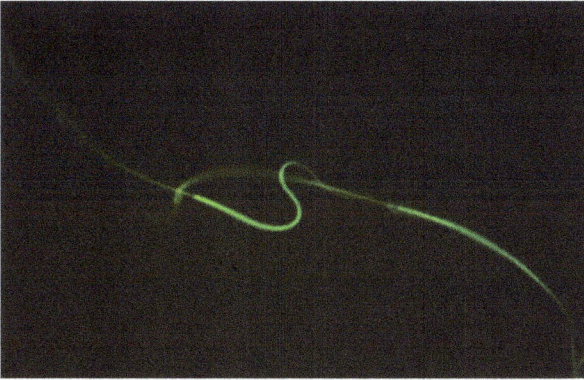

Fig. 8.2: FITC fluorescence immunostaining (indirect method) shows the sperm associated antigen 11c at the main section of the sperm tail.

8.2.5 The elimination of nonspecific fluorescence

The generation theory of the nonspecific fluorescence in the immunofluorescence staining is very complicated; thus, it is quite necessary to find the reason and take proper methods to eliminate accordingly.

8.2.5.1 Source of the nonspecific fluorescence
a. Free fluorescence exists in fluorescence-labeled antibody, and it cannot be removed by general purification methods such as dialysis method because this part of the fluorescence can often form a comparatively huge polymer.
b. Combining with certain contents of the tissue cells, the nonspecific antibody or serum protein in the antiserum can be labeled with fluorescence in the preparation of fluorescent antibody.
c. Tissue cells may contain generic antigen, or cross-reactive antigens, which can combine with fluorescent antibody.
d. Excessively labeled antibody molecules have too much fluorescence molecules or large number of negative ions, and they can combine nonspecifically with normal tissues.
e. Tissue cells have spontaneous fluorescence.
f. Fluorescence is not pure, and the fixation of the tissue specimens is not proper.

8.2.5.2 Methods of eliminating nonspecific fluorescence
This usually includes absorption method of animal organs powder or homogenate, dialysis method, chromatography, fluorescence antibody dilution method and Evans blue staining.

8.3 Immunoenzyme method

Immunoenzyme technology appeared later than the immunofluorescence method, but it develops rapidly and has been widely used.

8.3.1 Theory

The core of the immunoenzyme technology is to label the antigen-antibody reaction with enzymes (mostly horseradish peroxidase) and then to show the chemical composition of tissue cells with the help of enzyme histochemistry method.

8.3.2 Common methods

The indirect method is commonly used. Through covalent bond combination of the secondary antibody, enzymes can combine with antigen-first-antibody complex of the tissue biopsy. The final insoluble colored product of the reaction is formed after the special substrate is catalyzed through the enzyme histochemical reaction. This makes the qualitative, quantitative or positioning research of the components of the antigens of cells or tissues possible.

8.3.2.1 Staining procedures

a. The paraffin is removed from the paraffin sections and then immersed in the distilled water followed by routine methods; frozen sections are adhered to glass slides and dried then directly put into Tris buffer saline (TBS) or phosphate buffer saline (PBS).

b. Immerse the sections in 0.3% H_2O_2 (dissolve in double distilled water, TBS or methanol) for 30 minutes to seal the activity of the endogenous enzyme.

c. Rinse the sections twice in TBS, 5–10 minutes each time.

d. Immerse the sections in 0.2%–1% Triton X-100 or 0.1% saponin for 30 minutes.

e. Rinse the sections twice in TBS, 5–10 minutes each time.

f. Add 3% normal serum (from same origin of the secondary antibody animal), and incubate the sections for 30 minutes.

g. Discard the upper liquid, do not rinse and add the primary antibody directly; incubate for 24–48–72 hours at 4°C. Add 1% normal goat serum in the antibody diluent.

h. Rinse the sections twice in TBS, 5–10 minutes each time.

i. Mark the second antibody with enzyme 1:50–1:100 and incubate the sections at room temperature for 30–60 minutes.

j. Rinse the sections twice in TBS, 5–10 minutes each time.

k.   Stain with 0.05% diaminobenzidine (DAB)–0.01% H_2O_2 (prepared with 0.1 mol/L
     TB, pH 7.6) for 2–5 minutes, and control the staining density under microscope.
l.   Rinse the sections in TBS and double distilled water.
m.   Dehydrate, clear and then mount sections with neutral gums.

Note: If the activity of endogenous enzymes or antibodies penetration can be ensured,
steps b–d can be omitted.

8.3.2.2 Evaluation

The final products of reaction can be observed with an ordinary microscope with fine
structure details; it is not essential to use a fluorescence microscope. Specimens can be
preserved for a long time. Comparatively high electron density can be created by osmium
oxidation of DAB, and this is the reason why it can be observed with an electron microscope. However, the activity of the antibody and the enzyme may be damaged because of
the covalent binding between them. Meanwhile, the nonspecific antibody in the antiserum
is also marked by the enzyme; thus, it is not used as widely as the PAP method.

8.3.3 PAP method

The PAP method is one of the most commonly used immunohistochemical methods.
The antienzyme antibodies of PAP complex and the primary antibodies are from
the same animal so the secondary antibody can be used as the bridge antibody of
combining antibody to connect the PAP complex and the primary antibody, thus
leading the enzyme to the antigen to form the colored final product through enzyme-
histochemical reaction (Fig. 8.3).

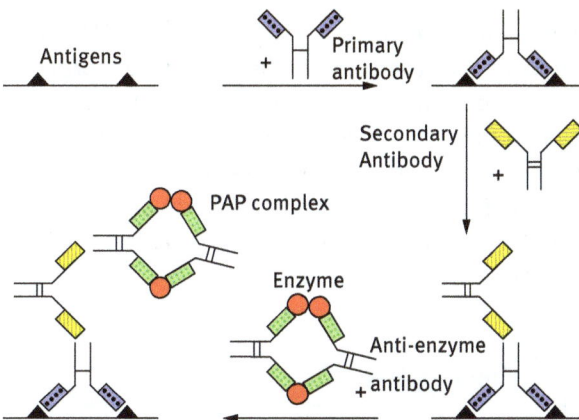

Fig. 8.3: Basic theory of the PAP method.

8.3.3.1 Staining procedure

a. The paraffin should be removed from the paraffin sections and then immersed in the distilled water. Frozen or vibration sections can stick directly onto the slide. They can be dried and then rinsed in TBS or dyed by floating in it because TBS can remove the fixative.
b. Immerse the sections in 0.3% H_2O_2 (made with the double distilled water, TBS or pure methanol) for 10–30 minutes.
c. Rinse the sections twice in TBS, 5–10 minutes each time.
d. Immerse the sections in 0.2%–1% Triton X-100 or 0.1% saponin for 30 minutes.
e. Rinse the sections twice in TBS, 5–10 minutes each time.
f. Drop 3% normal serum (for example, goat serum), and incubate the sections for 30 minutes; discard the serum without rinse.
g. Incubate the first antibody at 4°C for 24–72 hours and place in the wet box; floating dyeing with 24-hole culture plate sealed with the cover.
h. Rinse the sections twice in TBS, 5–10 minutes each time.
i. Incubate the second antibody (such as goat rabbit IgG resistance) 1:100 at room temperature for 30 minutes.
j. Rinse the sections twice in TBS, 5–10 minutes each time.
k. Incubate the rabbit PAP complex 1:100 at room temperature for 30 minutes.
l. Rinse the sections twice in TBS, 5–10 minutes each time.
m. Stain with 0.05% DAB–0.01% H_2O_2 (prepared with 0.1 mol/L TB, pH 7.6) for 2–5 minutes; control the staining density under microscope.
n. Rinse the sections in TBS and double distilled water.
o. Dehydrate, clear and mount sections with neutral gums or glycerin gel.

8.3.3.2 Notes

The primary antibody should be diluted as much as possible when the PAP staining method is implied. Bigbee thought that the distance between neighboring Fc region combining with the antigen can be shortened because of the overdosage of the primary antibody and the abundance of tissue antigens. If the shortened distance is suitable to the distance between two Fab regions of the bridge antibody, the two Fab regions of the bridge antibody will combine with the primary antibody, and the result shows negative because they cannot combine with the PAP complex. Thus, in this way, false-negative results may occur (Fig. 8.4).

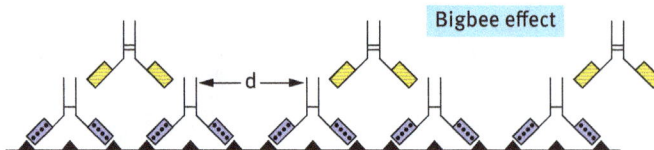

Fig. 8.4: Bigbee phenomenon in the PAP method.

High concentration of the secondary antibody should be ensured to guarantee that the secondary antibody (the bridge antibody) can combine with the primary antibody with only one Fab region and the other Fab region will be free to combine with the antienzyme antibody of the PAP complex.

8.3.3.3 Advantages of the PAP method
It has high specificity because all antibodies are not marked to avoid the covalent binding between the antibody and the enzyme and to maximize the activity of the antibody and enzyme, thereby enhancing the specificity of staining.

It has high sensitivity. The multilayer antigen-antibody reactions amplify the reaction, and simultaneously the PAP complex is very stable. This can increase the enzyme molecules on the antigen sites to enhance the specific staining, and the specificity is also improved.

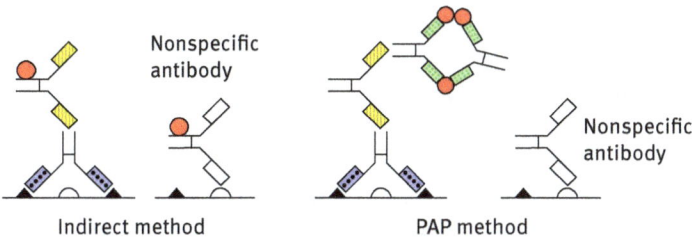

Fig. 8.5: Source of background staining of the indirect method and reason of light background staining of the PAP method.

Even if the bridge antibody has nonspecific antibody, it cannot connect rabbit PAP complex (Fig. 8.5) rabbit IgG antibody because it is not antirabbit antibody. Thus, nonspecific staining will not occur by bridge antibody itself.

The primary antibody in the PAP method is diluted greatly, and this lowers the background staining and increase the signal-to-noise ratio.

8.4 Avidin-biotin method

Avidin and biotin were applied to the immunohistochemistry in the 1970s–1980s. Their high affinity and immediate reaction make the immunohistochemistry with an avidin-biotin system basis called affinity histochemistry. The avidin-biotin complex (ABC) method and the Streptavidin perioxidase (SP) method are commonly used at present.

8.4.1 Theory

The theoretical basis of the application of avidin-biotin system in immunohistochemistry includes the following reasons: (1) Avidin has a very high affinity with biotin (dissociation coefficient of 10^{-15} mol/L), and their combination can form stable and long-lasting complex because there are four biotin combination sites for one single avidin molecule. (2) Biotin can label macromolecular substances, such as antibodies, enzymes, etc. (3) Avidin can be marked with many markers such as enzymes, colloidal gold, fluorescent pigment, etc. (4) Avidin can be used as a bridge to connect two different biotin-labeled molecules, such as antibodies and enzymes. Thus, on the basis of antigen-antibody reaction, the purpose of inspecting antigens of tissue cells can be realized through the high affinity between avidin and biotin.

1. The ABC method. The primary antibody is used first, and then the biotin-labeled secondary antibody is applied. Next comes the ABC complex, and the final step is the enzyme histochemistry staining.
2. Labeled avidin-biotin (LAB) method. The method uses biotin to label the primary antibody (direct method) or the secondary antibody (indirect method), and it uses substances such as enzymes, fluorescence or colloidal gold to label avidin. The above-mentioned reagents are added as the previously mentioned procedures. Ordinary microscopes, fluorescence microscopes or electron microscopes can all be used for observation. Moreover, the indirect method is comparatively common. The SP method used at present is to label the streptomycin avidin with peroxidase enzyme.
3. Bridged avidin-biotin (BRAB) method and indirect bridged avidin-biotin (IBRAB) method. These two technologies label the primary antibody (BRAB method) and the secondary antibody (IBRAB method) with biotin, respectively. At the same time, the enzyme is labeled with biotin, and the unlabeled avidin is used as the bridge to connect the biotin-labeled antibodies and enzymes. The method is limited to the immunoenzyme method because biotin cannot label fluorescence or colloidal gold. To check the antigen, the primary antibodies are used to incubate slide (IBRAB) and rinsed away by TBS. Then biotin-labeled secondary antibodies are used and rinsed by TBS. Avidin is added, and the biotin-labeled enzyme is added in the final step and visualized by enzyme histochemistry technology.

8.4.2 ABC method

The characteristics of the ABC method include that the primary antibody is not marked and that the secondary antibody is marked by biotin. In addition, ABC complex is previously prepared by mixing avidin- and biotin-labeled peroxidase according to proper proportion to make sure that at least one of the four binding sites of the avidin molecule surface is free to combine with the biotin in the biotin-labeled antibody to form antigen-antibody complex with enzyme molecules and finally localized by coloration (Fig. 8.6).

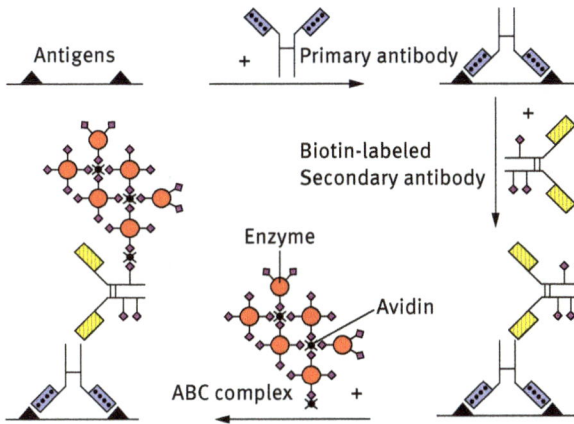

Fig. 8.6: Basic theory of the ABC method.

8.4.2.1 Staining procedures

a. The paraffin sections are dewaxed and then immersed in distilled water. Frozen or vibration sections can be adhered onto the slide. They can be dried and then rinsed in TBS or dyed by floating in it because TBS can remove the fixative.

b. Immerse in 0.3% H_2O_2 (dissolved in water, buffer or methanol) for 15–30 minutes.

c. Rinse the sections twice in TBS, 5–10 minutes each time.

d. Immerse in 0.2%–1% Triton X-100 or 0.1% saponin for 30 minutes.

e. Rinse the sections twice in TBS, 5–10 minutes each time.

f. Incubate the section in 3% normal goat serum for 30 minutes, then discard the serum; do not wash.

g. The first antibody (high dilution is acceptable) is introduced at room temperature for 30 minutes or at 4°C for the night.

h. Rinse the sections twice in TBS, 5–10 minutes each time.

i. The second antibody marked by biotin is introduced (1:200) at room temperature for 30 minutes.

j. Rinse the sections twice in TBS, 5–10 minutes each time.

k. The ABC compound (1:100) is introduced at room temperature for 30–60 minutes.

l. Rinse the sections twice in TBS, 5–10 minutes each time.

m. Stain with 0.05% DAB–0.01% H_2O_2 for 1–5 minutes, and control the degree of staining under a microscope.

n. Rinse the sections in TBS and double distilled water.

o. Seal the section routinely and observe (Fig. 8.7)

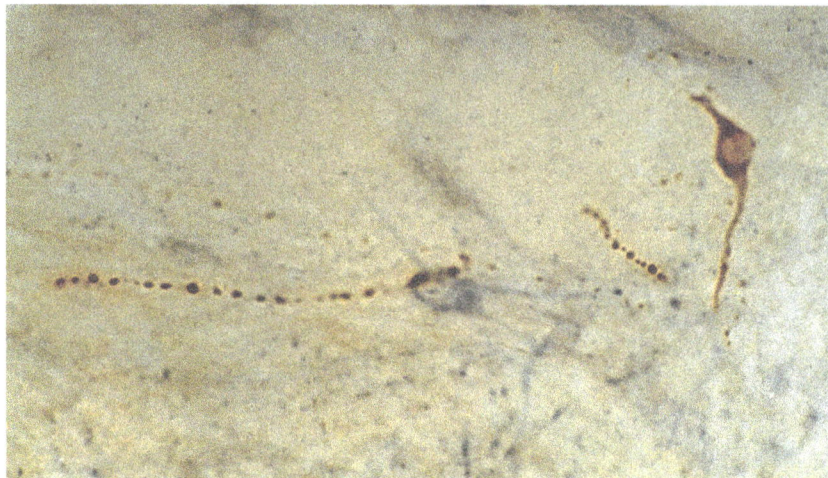

Fig. 8.7: ABC method shows the GnRH-positive neurons in hypothalamus.

8.4.2.2 Notes

Endogenous avidin or biotin combining sites should be blocked. Certain tissues and cells, such as liver, kidney, fat tissue, breast glands and white blood cells, are rich in enzymes containing biotin. When the materials are the previously mentioned tissues or cells, the following methods can be used to block the combining activity of the tissues or cells after eliminating the endogenous enzyme activity:

a. Incubate with 0.01% avidin for 20 minutes.
b. Rinse the sections twice in TBS, 5–10 minutes each time.
c. Incubate with 0.01% biotin for 20 minutes.
d. Rinse the sections twice in TBS, 5–10 minutes each time.

The combining sites of the tissue nonspecific avidin also need to be noted. Because avidin contains sugar residues, including mannose and glucosamine, these residues can react with the lectinoid protein in the tissues or adhere to certain tissue components. Avidin has high isoelectric point (PI, 10.5); thus, it can combine with the negative-charged tissue parts to create nonspecific staining. Streptavidin can be used to replace avidin in the streptavidin-ABC (SABC) method. Streptavidin is abstracted from streptomyces, but it has no sugar residues, and the PI is close to neutral (6.5). Thus, this method is widely applied at present.

The particles of mast cells can combine with avidin, and it should be replaced by streptavidin when necessary.

The reagents in different bottles in one ABC kit should be used with the reagents in the same kit. ABC compound should be prepared 30 minutes before using, and the preparation should strictly follow the instruction.

If the adhesive contains egg white containing avidin, the ABC method is not proper.

The advantages of the ABC method are as follows:

a.   The ABC method has high sensitivity. The antibody molecule marked by biotin has combined many biotin molecules, and each biotin molecule can combine with the ABC compound to increase the number of enzyme molecules.

b.   Background staining is slight in the ABC method. Because the ABC method is very sensitive, the primary and the secondary antibodies can be diluted to a very high ratio; hence, the nonspecific background staining is greatly decreased.

c.   The method is simple and convenient. The ABC method does not need the preparation of antienzyme antibodies. Like the PAP method, it is one of the most commonly used immunohistochemistry methods.

8.4.3 SP method

In recent years, the SP method is widely applied because of the huge supply of kits and their easy operations. The SP method is, in fact, the indirect method of the LAB method to use the secondary antibody marked by biotin and streptavidin labeled by peroxidase to detect the antigens of the tissue cells.

The staining procedures are as follows:

a.   The paraffin should be removed from the paraffin sections and then immersed in the distilled water. Rinse the sections three times in PBS, 5 minutes each time.

b.   Each section is added with endogenous enzymes blockers (reagent A, 50 μL) at room temperature for 10 minutes.

c.   Rinse the sections three times in PBS, 5 minutes each time.

d.   Seal the serum (reagent B, 50 μL) at room temperature for 10 minutes. Then discard the serum; do not wash.

e.   Incubate the primary antibody (50 μL; high dilution is acceptable) at room temperature for 60 minutes or 4°C for the night.

f.   Rinse the sections three times in PBS, 5 minutes each time.

g.   Incubate the secondary antibody marked by biotin (reagent C, 50 μL) at room temperature for 10 minutes.

h.   Rinse the sections three times in PBS, 5 minutes each time.

i.   Add enzyme-marked streptavidin (reagent D, 50 μL) at room temperature for 10 minutes.

j.   Rinse the sections three times in PBS, 5 minutes each time.

k.   Stain with 0.05% DAB−0.01% H_2O_2 for 1–5 minutes.

l.   Rinse the sections in TBS and double distilled water.

m.   Seal the section routinely.

8.5 Protein A method

Protein A is the abbreviation of staphylococcus protein A. Protein A is the protein separated from the cell walls of *Staphylococcus aureus*.

8.5.1 Nature of the protein A and its applications

8.5.1.1 Introduction of protein A
Protein A is composed of single polypeptide chains. Its molecular mass differs if the extraction methods vary, ranging from 12 to 42 kDa, and its electric point is 5.1. Protein A has a wide variety of resource, and it is economical to be applied in experiments for it can combine with many immunoglobulins of mammals and has no special requirement of the species. The combination happens in the Fc period, and this will not influence the activity of the antibody. In addition, protein A can be marked by many markers such as fluorescence, enzymes and colloidal gold.

8.5.1.2 Application of protein A in immunohistochemistry
Protein A has a bivalent binding force and, thus, can be used as a bridge antibody. It can also be marked and used as a marker to label antibody. Thus, protein A can replace the secondary antibody in immunohistochemical staining. Protein A is small in molecular weight; hence, it can easily penetrate tissues.

8.5.2 Staining method

Enzyme-marked protein A is used in the indirect method.
a. The sections can be processed as previously mentioned. Rinse the sections in TBS.
b. Immerse the sections in 0.3% H_2O_2 for 30 minutes to block the activity of endogenous enzymes.
c. Rinse the sections twice in TBS, 5–10 minutes each time.
d. Immerse the sections in 1% ovalbumin (prepared with TBS) for 20 minutes, discard and do not wash.
e. Immerse the sections in the primary antibody at room temperature for 30 minutes, or at 4°C overnight.
f. Rinse the sections twice in TBS, 5–10 minutes each time.
g. Treat the sections with enzyme-labeled protein A 1:100 to 1:400, 30 minutes.
h. Rinse the sections twice in TBS, 5–10 minutes each time.
i. Use DAB-H_2O_2 treatment as usual.
j. Rinse the sections in TBS and double distilled water.
k. Seal the sections.

Note: Before the primary antibody is introduced, the ovalbumin instead of normal serum is used to block the nonspecific binding sites because protein A can combine with normal serum IgG.

Protein A's application in the PAP method is as follows:

a. The sections can be processed as previously mentioned. Rinse the sections in TBS.
b. Immerse the sections in 0.3% H_2O_2 for 30 minutes to block the activity of endogenous enzymes.
c. Rinse the sections twice in TBS, 5–10 minutes each time.
d. Immerse the sections in 1% ovalbumin (prepared with TBS) for 20 minutes, discard and do not wash.
e. Immerse the sections in the primary antibody, at room temperature for 30 minutes, or at 4°C overnight.
f. Rinse the sections twice in TBS, 5–10 minutes each time.
g. Immerse in protein A (1 µg /mL) for 15–30 minutes.
h. Rinse the sections twice in TBS, 5–10 minutes each time.
i. Immerse in PAP complex (no species limit) for 15–30 minutes.
j. Rinse the sections twice in TBS, 5–10 minutes each time.
k. Seal the section.
l. Use colloidal gold mark protein A method (discussed later).

8.6 Immunogold and immunogold-silver method

Colloidal gold was first introduced into the immunohistochemistry in 1971 and was mainly used to check antigens of cell surface with an electron microscope. In 1978, Geoghegan and his colleagues first used colloidal gold technology in optical microscope observation to check B-lymphocyte surface antigen. Colloidal gold particles larger than 10 nm cluster together appear red under bright field microscopes. This chapter mainly discusses colloidal gold method under optical microscopes, including immunogold method and immunogold-silver method.

8.6.1 Immune colloidal gold method

8.6.1.1 Theory

The basis of IGS is the stable and irreversible complex formed by the colloidal gold particles attached to big molecule mass such as antibodies and protein A. The activity of the marked antibody and protein A is not influenced because this attachment is noncovalent bond. The indirect method is frequently used in IGS, which means adding the primary antibody in preparation of tissues and cells followed by adding the gold-marked secondary antibody or gold-marked protein A. Then it can be observed directly under a microscope. The combining sites of the colloidal gold show pink, and the gold particles look bright golden spots under a dark field microscope. IGS-stained samples can be dehydrated and preserved forever.

8.6.1.2 Staining procedures

a. The paraffin should be removed from the paraffin sections and then immersed in the distilled water. Rinse the sections in 0.1 mol/L TBS (pH 7.4).
b. Drop 3% normal goat serum and incubate the sections for 30 minutes (1% ovalbumin used in protein A method), discard the serum and do not wash.
c. Immerse in the primary antibody at room temperature for 120 minutes, or at 4°C for the night.
d. Rinse the sections two times in 0.1 mol/L TBS, pH 7.4.
e. Rinse the sections in 0.1% bovine serum albumin (BSA)–20 mmol/L TBS (pH 8.2) for 5–10 minutes (pH 7.4 in protein A method).
f. Immerse in 3% normal goat serum (at pH 8.2 BSA-TBS configuration) for 30 minutes (BSA containing 1% ovalbumin TBS, pH 7.4, preparation in the protein A method).
g. Immerse the colloidal gold-labeled second antibody (or protein A) at room temperature for 60–120 minutes, or at 4°C overnight. (The staining can be controlled under microscope.)
h. Rinse the sections twice in BSA-TBS.
i. Rinse the sections in double distilled water.
j. Perform postfixation with 1% glutaraldehyde for 10 minutes.
k. Rinse the sections in double distilled water.
l. Dehydrate, clear and mount.

8.6.1.3 Notes

a. Gold particle size selection: 30–40 nm diameter to indicate the cell surface antigen; 10–20 nm for organization intracellular antigens.
b. To increase the gold particles penetration: Penetration is poor because of the large size of gold particles; thus, a detergent of 0.2%–1% Triton X-100 is normally used to proceed the sections (unnecessary for the procession of cell surface antigen).
c. Aldehyde fixed organization: 1% sodium borohydride, 20 mmol/L lysine or 0.2–0.5 mol/L NH_4Cl is used to precede the sections to remove the free aldehyde.
d. Increase the visibility of the gold particles: IGS is not the most sensitive method, and this is usually caused by the fact that tissue antigen is rare, ideal dyeing is not easy to obtain usually and it is not clear in coloration of nerve fibers. Thus, IGS often needs a relatively high concentration of colloidal gold probe, which limits the application of IGS method in the display of antigens of biopsy.

If necessary, the following methods can be used to increase the visibility of the gold particles:

i. After being stained, the sections of colloidal gold-marked secondary antibody can be processed with gold-marked IgG (from the same species as the first antibody animals). If the pAg method is applied, anti-protein A antibody and pAg can be added after adding pAg. Through this processing, the number of gold particles of the antigen sites can be increased to amplify the staining results.

ii. Photography can be done with dark field microscopes, but cell morphology observation should be conducted under a bright field illumination.

iii. Use silver strengthening method (see next section).

8.6.2 Immunogold-silver method

Immunogold-silver staining (IGSS) is established based on IGS. After dyeing, images appear in the solution containing silver ions and reducing agent. The silver ion is reduced to silver particle adhering around the colloidal gold particles by the action of the reducer, and the gold particles act as the catalyst. Then by working as the catalyst, the silver articles reduce many silver ions to increase the size the silver shell around the colloidal gold particles to make the former invisible gold particles much more visible and show brown black to black under an optical microscope (Fig. 8.8).

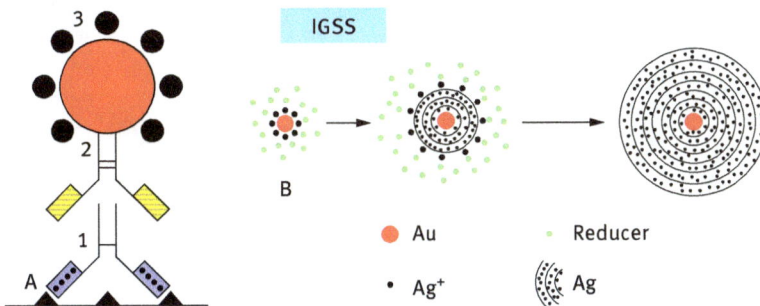

Fig. 8.8: Basic theory of IGSS.

Review question

What should we need to pay intense attention if we use mouse-raised primary antibody to stain the samples from mouse?

9 Specificity and sensitivity of immunohistochemistry

Immunohistochemistry uses the specific antigen-antibody reaction to detect the substances with antigenicity. The end result achieved by immunohistochemistry methods depends on the final deposition, which results in two questions: How do we make sure the staining results are based on the antigen-antibody reaction? The preparation of the organization unavoidably leads to the loss of a few or a lot of antigen, how do we detect the trace antigen in the tissue or cells using the immunohistochemical method? The main point is the immunohistochemistry specificity and sensitivity, which mainly depends on the primary antibody.

9.1 Specificity and immunohistochemistry staining

9.1.1 Specificity

The specificity of immunohistochemistry includes two main concerns: the specific method and the specific serum antibody. The specific method means that the immunohistochemistry staining results come from the reaction between the specific antibody and the corresponding antigen. Specific serum antibody means that the staining results are based on the reactions between the detected antigen and antibody, which proves that the exact antigen is revealed; meanwhile, the cross-reaction is eliminated.

9.1.2 Specific and nonspecific staining

The process of immunohistochemical staining is always accompanied by a certain background color or a nonspecific staining. Thus, distinguishing specific staining from nonspecific staining is very important.

9.1.2.1 Specific staining
Specific staining means the specific immunological response is caused by the primary antibody and the corresponding antigen. The combined products are often distributed in a particular area and reflect a certain structure, i.e., the cytoplasm, the cell surface, the nucleus or a special structure within the organization. Even on the same tissue slice, because of the different cell cycles, the staining density could be different. In addition, to confirm the specific staining, the experiment should be repeated and a variety of preparation process (such as frozen sections, paraffin, etc.) or various staining methods (e.g., immunofluorescence, enzyme immune, etc.) should be attempted, especially for the newly discovered positive results.

DOI 10.1515/9783110531398-009

9.1.2.2 Nonspecific staining

The nonspecific staining of tissue cells means that the immunological response is not caused by the primary antibody and the antigen. Nonspecific staining results normally appear as diffused or disorder distribution and without a good morphological distribution. Sometimes both cells and its surrounding structures or the nuclei of all cells and some unexpected structures are stained simultaneously. Normally, the results of several experiments conflict with one another. Nonspecific staining of tissue sections is also common in the edge, particularly the edge of the drying sections (called edge effect) or with folding marks, bad fixed tissue center and organized necrotic area. Excessive nonspecific staining will disturb the observation and record of specific staining results, and it will affect the quality of the micrograph.

9.2 Control experiment

During the process of immunohistochemical staining, specific immunological tests mainly depend on a series of control experiments. Any immunohistochemical staining process must include certain control experiments to demonstrate the specificity of the experimental results and exclude nonspecific staining. Experimental results without controlled trials are unconvincing. The control experiments generally include positive and negative controls.

9.2.1 Positive controls

Positive controls are designed to demonstrate that the staining method and all reagents used are valid. If negative reactions occur in the positive control, it is called false negative. The possible causes and treatment when false-negative reaction appears are as follows:

a. During the preparation process of cells and staining, the soluble antigens are lost from the specimen. An effective fixing prevents the loss of antigen.
b. In the process of fixing and embedding, antigen epitopes may be screened or changed, thus weakening its immunological response. That could be solved by changing the fixation and embedding processes such as fixation with aldehyde or an aldehyde-containing fixative, or by rinsing thoroughly after fixation and trying the protease digestion to release crosslinks in antigen molecules.
c. When the thick sections especially cut with vibration or freezing microtomes are used, incubation of the sections with the antibody is not in contact with the antigen. Choose the way to increase antibody penetration.
d. Staining method problems. The method is less sensitive, so very little of the antigen cannot be detected. Alternatively, the PAP method is introduced and the

antigen content is rich in tissue sample. The concentration of primary antibody is too high; hence, all the two Fab fragments of the bridge antibody combine with the Fc fragments of the primary antibody, and the PAP complexes could not connected with the bridge antibodies, resulting in the false negative. If this happens, a highly diluted primary antibody is recommended.

e. Antibody titer is too low. The reason may be the antibody obtained from the serum animals is not high, or the antiserum undergoes prolonged storage or repeated freezing and thawing. The solution is to replace this one with a new primary antibody. Sometimes secondary and tertiary antibodies have the similar situation.

9.2.2 Negative controls

Negative controls generally include the blank test, the substitution test, the blockade test and the absorption test. Negative control test results should be negative. If the result is positive, it must be false positive, then try to identify the causes of false positives and eliminate all of them.

9.2.2.1 Blank test
By incubating with the antibody dilution buffer (phosphate-buffered saline [PBS], Tris-buffered saline [TBS], etc.) instead of the first antibody in the staining procedure, the result should be negative.

9.2.2.2 Substitution test
The same concentration of normal serum (from the same animal species in which the primary antibody is prepared) is used instead of the primary antibody. Because of lacking specific antibodies, the result should be negative. When using a polyclonal antibody (antiserum), the normal serum must be used as a negative control test. If the antiserum contains other substances that could cause nonspecific staining, then the normal serum is used as a control to discover the nonspecific reaction. On the contrary, when using PBS or TBS, in which nonspecific staining does not occur, the result will be regarded as specific staining.

9.2.2.3 Blockade test
Tissue sections are immersed in the buffer with excess of unlabeled antibodies first then incubated with labeled specific antibodies. Because antigens have combined with unlabeled antibodies preferentially rather than labeled ones, the result should be negative. And when performing immunostaining with the mixture of excess of unlabeled antibodies and labeled specific ones, the result is also negative.

9.2.2.4 Absorption test

Absorption test is the most effective control test, particularly when a new type of antibody or organization has not checked previously. Excess antigen is used to absorb the corresponding specific primary antibody in the antiserum. Then the antiserum is introduced in the staining procedures after absorption. The result should be negative.

9.3 Methods to enhance the immunohistochemical sensitivity

The purpose of immunohistochemistry is to obtain clear positive results with very little or no background staining. This can be obtained by enhancing specific staining and/or reducing nonspecific background staining.

9.3.1 Enhanced specific staining

9.3.1.1 Select sensitive method

PAP method, ABC method, immunogold or immunogold-silver labeling method could be selected to increase the number of markers for tissue antigens and then the amount of immune reaction products.

9.3.1.2 Increase antibody incubation time

At high dilution, prolonged incubation time at 4°C increases the balance of the primary antibody, the antibody easily reaches the tissue antigen. Because the primary antibody is diluted, high background staining is also reduced. Furthermore, the incubation time of secondary antibody or the third reagent layer can also be prolonged.

9.3.1.3 Repeat antibodies or detection reagents

9.3.1.3.1 Repetition of the primary antibody

The tissue slides are thoroughly cleaned after the primary antibody incubation. Thus, the first low-affinity antibodies after incubation are rinsed, leaving the binding sites of antigens to bind the primary antibodies in the second treatment, thereby increasing the intensity of staining.

9.3.1.3.2 Repetition of the secondary antibody and the PAP complex

In the PAP method, the step of the secondary antibody with PAP complexes is repeated again after combining with the PAP complex and the final coloration. Repeated bridge antibody combines with the unsaturated antigen combining sites on the first application of antibodies (Fig. 9.1A and B) or the first antibody (Fig. 9.1C). Then the connection of the PAP complex is repeated, resulting in a huge presence of complex containing more enzyme molecules to enhance staining results.

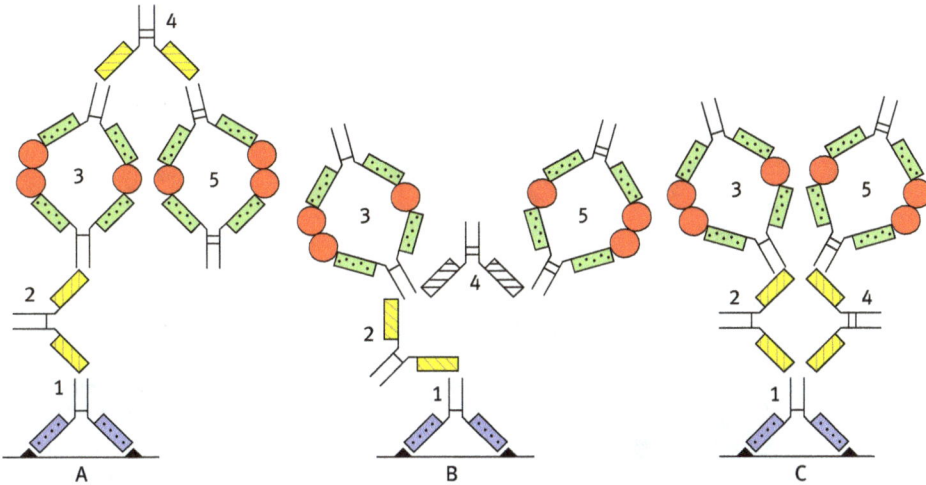

Fig. 9.1: Possible amplification mechanism for repeated secondary antibody and PAP complex.

9.3.1.4 Protease digestion and antigen retrieval (see Chapter 7)

9.3.1.5 Increasing the penetration of antibodies and detection reagents (see Chapter 7)

9.3.1.6 Enhancing the coloring of DAB reaction product
The following methods can be used to strengthen the coloring of 3,3′-diaminobenzidine (DAB) reaction products.

9.3.1.6.1 Method of O_SO_4 enhancement
After the DAB reaction, the sections react with 0.01%–1% O_SO_4 solution (dissolved in TBS or PBS) for 30 seconds or longer. The production of the osmium is brownish black and can also be observed under electron microscopy.

9.3.1.6.2 Method of nickel enhancement
Add 8% of 50 μL $NiCl_2$ into 10 mL of DAB-H_2O_2 solution. DAB reaction product is purple blue when there is a presence of Ni^{2+}.

9.3.1.6.3 Method of cobalt and nickel enhancement

DAB (100 mg) is dissolved in 200 mL of 0.1 mol/L PB with pH 7.3, then 5 mL of 1% cobalt chloride and 4 mL of 1% nickel ammonium sulfate are added. The sections are incubated for 10 to 15 minutes, and the reaction product is black.

9.3.1.6.4 Method of imidazole enhancement

Imidazole (10 mmol/L) in DAB solution results in brown-red product (Fig. 9.2).

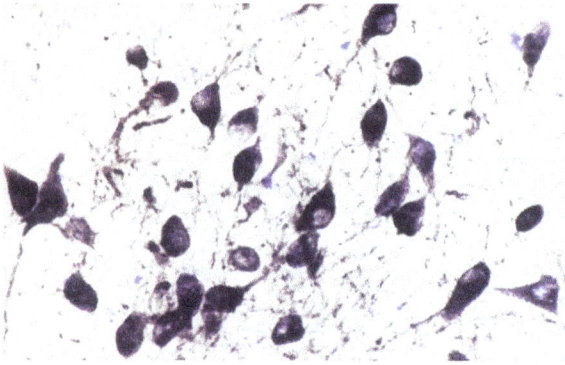

Fig. 9.2: ABC method, DAB-H_2O_2-nickel ammonium sulfate color. The positive neurons in mice midbrain nigra tyrosine hydroxylase.

9.3.2 Causes of background staining and the methods of elimination

Nonspecific staining often interferes with specific staining. The following methods can reduce nonspecific staining and improve the sensitivity.

9.3.2.1 Fixation inadequate or inappropriate

Inadequate or inappropriate fixation releases the cell antigen into the extracellular tissue and causes background staining. Inappropriate fixatives need to be replaced by better fixatives. For example, the cacodylate buffer easily causes the background color; it can be replaced with other buffer to prepare the fixative solution.

9.3.2.2 Section drying during the processes

During the dyeing process, if the solution evaporates and the tissue slice becomes dry, it may cause serious nonspecific staining. Sometimes a wet box is used in immuno-histochemical staining, especially when the sections are incubated with antibody for a longer time at 37°C.

9.3.2.3 Endogenous activity

Endogenous enzymes include peroxidase in the brain tissue, macrophages, granu-locytes and other tissues or iron porphyrin in erythrocytes. The substrate could be oxidized by activities of these enzymes and lead to false-positive results, especially in frozen and vibration sections. Usually before the primary antibody is introduced, the sections are treated with 0.3% H_2O_2 in distilled water or a buffer solution or prepared in methanol for 15 to 30 minutes to inhibit endogenous enzymes activities. If H_2O_2 treatment destroys antigen, add H_2O_2 after the primary antibody incubation.

9.3.2.4 Free aldehyde
When aldehydes (paraformaldehyde and glutaraldehyde) fix tissue and the fixative solution is not completely rinsed before dyeing, a free aldehyde group will be left and nonspecific staining may occur. In this case, 1% sodium boron hydride (potassium), 20 mmol/L lysine or glycine, 0.1 mol/L NH_4Cl or normal serum albumin will be used to seal the free aldehyde groups.

9.3.2.5 Fc receptor
Normal animal serum from which the secondary antibody is prepared is used to blockade the Fc receptor. Fc binding sites are occupied by normal serum immunoglobulin, therefore preventing the combination of primary antibody. Moreover, the secondary antibody does not react with the blocking serum immunoglobulins because they are from the same species.

9.3.2.6 Hydrophobic bonds and ionic bonds
Immunoglobulin could combine with tissue proteins or other components through hydrophobic bonds or ionic bonds, and complement in the antiserum could also bind to tissue proteins and then combine with immunoglobulin.

The following methods can eliminate nonspecific combination:
a. Block with normal serum. Because of this low-affinity binding, the normal serum is removed without rinse after blockade, and then primary antibody is added directly.
b. Dilute primary antibody as much as possible.
c. Inactivate complement in antiserum (heated to 56°C, 30 minutes).
d. Increase the ionic concentration of the rinse liquid, such as using TBS (high ionic concentration compared with PBS), in which NaCl can be added to the 2.5% (0.5 mol/L).
e. Apply detergents such as Triton X-100, saponin, etc. (see Chapter 7).

9.3.2.7 Natural antibodies and antibody impurities
Natural antibody means that the animals have been exposed to certain substances that produce antibodies before they are immunized with specific antigens. Antibody impurity means that the impure antibodies cause certain contaminative antibodies components, and these antibodies may be present in the antiserum and cause nonspecific staining. Generally, the highly diluted antiserum can reduce the background staining.

9.3.2.8 Carrier protein antibody

During the preparation of small molecule peptide antibody, the carrier protein such as bovine serum albumin (BSA) may produce its own antibodies. Such carrier protein antibodies can bind to similar tissue protein, causing background staining. Carrier proteins can be used to absorb the antiserum to eliminate nonspecific staining.

9.3.2.9 Autofluorescence

Some tissues may have autofluorescence components, such as elastic fibers, collagen fibers, lipofuscin and fluorescence, that can be induced by aldehyde fixative solution. If this happens, 0.01% Evans blue-PBS will be used to dilute fluorescence-labeled antibody, or choose other immunohistochemical methods if such autofluorescence seriously affects the results.

9.3.2.10 Inherent pigmentation

The melanin in the neurons of substantia nigra makes it difficult to distinguish with the reaction products. Replace it with different electron donors or other methods.

9.3.2.11 Add protein into antibody dilution

Add 1% normal serum (from the same species in which the secondary antibody is produced) or BSA, and these additional proteins will combine with impurity antibodies in antiserum or with nonspecific binding sites, which could reduce the nonspecific staining.

9.3.2.12 Thorough cleaning

Thoroughly clean sections between each two steps of staining (except for the blockade with normal serum); the nonspecifically bound antibody or other agent can be removed to reduce background staining. Usually rinse with TBS or PBS two to three times for 3–5 minutes.

9.3.2.13 Cross-reactivity

Because it is truly an antigen-antibody reaction, it is difficult to deal with. If possible, try to use monoclonal antibody, or use antibodies from multiple sources. The results should be understood comprehensively.

Knowledge links

Ida CM, Vrana JA, Rodriguez FJ, et al. Immunohistochemistry is highly sensitive and specific for detection of BRAF V600E mutation in pleomorphic xanthoastrocytoma. Acta Neuropathol Commun. 2013;1(1):20.

10 Double-staining immunohistochemistry technology

Double-staining immunohistochemical technology allows two or more antigens to be revealed on the same section for discovering their locations, morphology and interrelations. Many double-staining immunohistochemical methods, which are suitable for light and electron microscopy, have been established. Multiple-staining immunohistochemistry has roughly the same basic principle as that in double-staining method.

10.1 Double-staining immunohistochemistry on serial sections

Double-staining immunohistochemistry on serial section is the simplest and securest double immunostaining. It reveals an antigen on two contiguous sections by the same immunohistochemical method, compares the location and composition of antigens that exist on these sections and examines whether they exist in the same cell or structure or not. The interference between two immunostaining results or the false double staining can be avoided efficaciously because each section is stained only once. The tissue blocks are cut into 1- to 3-μm sections; otherwise, the same cell or microstructure will not appear on two contiguous sections simultaneously. The thicker sections, such as nerve tissue (10–20 mm), used for this immunostaining method should adopt the "mirror" mounting way on the tissue slices to keep histological identity.

Double-staining immunohistochemistry on serial sections applies to the tissue slice containing some special structures with obvious morphological markers, small intestine and nervous tissue, for example. It is very difficult to identify the same cell in homogeneous tissue nevertheless. The other double-staining technology should be used to study the interrelations between an antigen and another one on the same section.

10.2 Immunofluorescence double-staining technology

Different kinds of fluorescence emit their own corresponding fluorochromes when they are excited; thus, they are used to label two or more antibodies. After immunohistochemical staining, two kinds of antigen material that individually bind to fluorescence-marked antibodies will show different colors and be detected clearly.

DOI 10.1515/9783110531398-010

10.2.1 Direct immunofluorescence double-staining technology

In direct labeling double-staining technology, two specific primary antibodies are conjugated directly to two kinds of fluorochrome, and the most popular combination fluoresceins are fluorescein isothiocyanate (FITC) and tetramethyl rhodamine isothiocyanate (TRITC) or CY2 and CY3. The sections can be incubated in mixtures of two primary antibodies according to optimal dilution (one step) or in one after another successively (two step). Observe the sections and take photos separately under fluorescent microscope with the correct filters to make sure that the wavelength of absorption and the emission of FITC and TRITC are correct and that the correlation between the two kinds of antigens are determined. To understand whether two antigens are present in the same cell, two or more photos with single fluorescence (green or red) can be immerged using some particular image analysis software. The portions where two antigens exist together will appear blend color, such as orange (green and red color mix). Although its sensitivity is poor, direct-labeling method is commonly used in immunofluorescence double staining because of the increase in commercial fluorescein-labeled monoclonal antibodies.

10.2.2 Indirect immunofluorescence double-staining technology

When performing indirect double staining, the most important criterion is that the two kinds of primary antibodies are not labeled and do not originate from different species of animals, such as rabbit and guinea pig. However, if monoclone antibodies are introduced, the antibodies may be different immunoglobulin (Ig) subtypes such as IgG and IgM or heavy chains from the same species animal. In this case, the two kinds of secondary antibodies may come from the same animal or different animal species (goat antirabbit IgG and goat anti-guinea pig IgG, or goat antirabbit IgG and pig anti-guinea pig IgG). They can be marked with FITC (or CY2) and TRITC (or CY3), respectively. In the staining procedures, the sections are incubated in the first kind of primary antibody and the corresponding first kind of fluorescence-labeled secondary antibody continuously to reveal the first kind of antigens. Then the sections are treated with the second kind of primary antibody and corresponding second kind of fluorescence-labeled secondary antibody to display the second kind of antigens. Of course, the primary and the secondary antibody incubation steps can be replaced by mixtures of two kinds of primary and then two kinds of fluorescence-labeled secondary antibodies, which can assist to simplify the protocol.

However, it has some specific disadvantages, as follows: (1) Four kinds of antibodies are used. (2) Two kinds of primary antibodies cannot originate from the same animals generally because of false double staining. The formaldehyde vapor treatment can be used for the indirect immunofluorescence double staining in which the two kinds of primary antibodies come from the same animal. After the first staining, formaldehyde vapor (80°C) is used to destroy residual antibody binding sites of this reaction and avoid false double staining. (3) Before the indirect double staining, interspecies cross test should be performed to identify the nonexistence of cross-reactivity, i.e., FITC-labeled goat antirabbit IgG does not react with guinea pig IgG, and TRITC-labeled pig anti-guinea pig also does not react with rabbit IgG.

Immunofluorescence double staining is the better choice to examine the coexistence of one trace antigen and another large number of antigens in the same cell (Fig. 10.1).

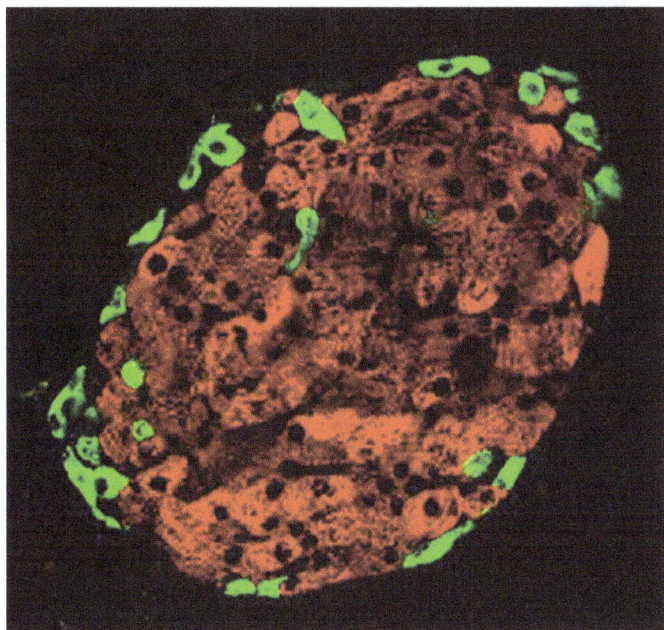

Fig. 10.1: Immunofluorescence double staining. Texas red-labeled secondary antibodies show insulin, and FITC-labeled secondary antibodies show glucagon. Mouse pancreas.

10.3 Immunoenzyme double-staining technology

Immunoenzyme double staining is a sensitive method and is widely used at present. Currently, one of the most popular immunoenzymes is indirect labeling or horseradish peroxidase-antiperoxidase (PAP) method in which a tertiary antibody is used.

10.3.1 Basic concept

In immunoenzyme double staining, the following processes should be understood.

10.3.1.1 Single-enzyme or double-enzyme labeling
This technology is classified into two subtypes: single-enzyme labeling and double-enzyme labeling.

a. Single-enzyme labeling An enzyme marker such as horseradish peroxidase (HRP) is used to reveal two kinds of antigens via different colorful final reaction products that are caused by different electron donors, for example, diaminobenzidine (DAB) and 4-chloro-1-naphthol (CN). This is mainly suitable for the immunoenzyme double staining in which two primary antibodies are derived from the same species. It is very important to remove all levels of antibodies and enzymes in the first antibody detection from the tissue sections between two staining, and this process is known as elution.

b. Double-enzyme labeling Two kinds of enzyme markers are used in this method. The step of elution is unnecessary, but primary antibodies must be originated from different species. The most popular enzyme markers are HRP and alkaline phosphatase (AKP) or HRP and glucose oxidase.

10.3.1.2 Optimum conditions of each single-labeling experiment

Before attempting a double-staining experiment, single-labeling experiment should be performed to find optimal dilutions for the different primary and secondary antibodies, reaction conditions and incubation time.

10.3.1.3 Sequence of double staining

The antigens, as well as the substrates upon which the enzyme acts, should be displayed in a specific order. Generally, the antigens that are less in tissue and are easily affected by staining should be stained first. DAB is commonly used as substrate solution in the first staining.

10.3.1.4 Strict controls

Performing antibody cross test is very essential because of the usage of two testing systems in double staining. Terminal conclusions of immunoenzyme double staining, which are obtained after contrasting with abundant and strict controls, are meaningful and convictive.

10.3.2 Antibody elution

Antibody elution, the key step for immunoenzyme double staining in which the primary antibodies derived from same species are applied, refers to the antibody residue need to be removed in the first staining from the sections. It must be performed completely, or there will be false double staining.

10.3.2.1 Possible reasons for false double staining

Taking PAP double-staining method, for instance, if not performing antibody elution, false double staining may occur. The possible reasons are as follows:

a. Immunological reactions occur between secondary antibody in the second staining and primary antibody in the first staining.

b. Immunological reactions occur between secondary antibody in the second staining and PAP complex in the first staining.
c. Immunological reactions occur between primary antibody in the second staining and secondary antibody in the first staining.
d. Immunological reactions occur between PAP complex in the second staining and secondary antibody in the first staining.
e. Enzyme of PAP complex in the first staining remains reactivity for the substrate in the second staining.

10.3.2.2 Common methods of antibody elution
Here we introduce three methods to get rid of antibody complex in the first staining.

10.3.2.2.1 Acid rinse
Antibodies may drop from tissue antigens according to the principle of antigen-antibody complex dissociation in acidic condition (pH 2.2).
a. Eluent glycine-HCl buffer, pH 2.2.

Glycine	0.75 g
NaCl	0.5 g
double distilled water (DDW)	100 mL

HCl (1 mol/L, approximately 5 mL) is gradually added to previously mentioned liquid to pH 2.2, or use 0.1 mol/L pH 2.0 HCl as eluent.
b. In the elution protocol, the sections are immersed in eluent 1–4 hours (time of elution is determined in reference to section thickness and antibody affinity) after the first staining. Replace eluent several times, rinse sections three times in Tris-buffered saline (TBS) then begin the second staining.

10.3.2.2.2 Antibodies on sections can be oxidized by acid plus KMnO$_4$
a. Oxidizing solution

2.5% KMnO$_4$	1 portion
5% H$_2$SO$_4$	1 portion
DDW	10–260 portions

b. Bleaching liquid, 0.5% Na$_2$S$_2$O$_5$
c. Elution protocol: Rinse sections in DDW after the first staining; add bleaching liquid until the color of tissue becomes white completely. Rinse several times in DDW and TBS successively and then begin the second staining.

10.3.2.2.3 Formaldehyde vapor method
In reference to Wang's formaldehyde vapor method, place 3 g of paraformaldehyde powder and the stained-once sections (or nickel net carried immunostained ultrathin

sections) into a 1-L container, and then keep the sections in 80°C oven for 1–4 hours. Rinse sections or nickel net in buffer, and continue to the second staining.

10.3.2.3 Check the effect of antibody elution

Checking whether antibodies are eluted completely or not is very important, same as the antibody elution process itself. The following methods can be used to examine the effect of all levels of antibody and enzyme elution.

a. Perform the first staining without color development (such as omitting DAB-H_2O_2 reaction).
b. Perform antibody elution.
c. Eluted sections can be divided into three groups:
    i.   First group: The chromogenic agent (such as DAB-H_2O_2) is added to the sections.
    ii. Second group: The PAP complex and the chromogenic agent are added to the sections.
    iii. Third group: The secondary antibody, the PAP complex and the chromogenic agent are added to the sections.
d. Result explanations
    i.   Negative result occurs in the first group: The PAP complex (including enzyme) is removed.
    ii. Negative result occurs in the second group: The secondary antibodies are removed.
    iii. Positive result occurs in the second group, but the first group is negative: The secondary antibodies are remained.
    iv. Negative result occurs in the third group: The primary antibodies are removed.
    v.   Positive result occurs in the third group, but the first and the second groups are negative: The primary antibodies are remained.
    Try alternate elution conditions according to results.
e. After elution treatment, perform the second staining on the unstained sections and then observe results: If the results are positive, the elution method has no effect on the revelation of the second antigen. If the results are negative, the elution method cannot be used because the second antigen has been destroyed.

In addition, the following procedures are also recommended: Carry on antibody elution after the chromogenic reaction of the first staining. Perform the second staining but do not add the primary antibody or its substitutes such as normal serum and TBS buffer, and then observe the results: If the results only contain some color of the second donor or mixed color, the antibody elution is performed completely.

10.3.3 Immunohistochemistry double staining with primary antibodies from different species

10.3.3.1 Basic theory
There are different kinds of enzymes applied in this method. The frequently used method is immunoenzyme double staining.

10.3.3.2 Steps of protocol
The PAP and the alkaline phosphatase-antialkaline phosphatase (APAAP) methods are introduced as follows:
a. Treatment of sections, inhibition of endogenous enzyme and block by normal serum are performed as previously described.
b. Incubate the sections for 30 minutes at room temperature or overnight at 4°C in the mixture of two primary antibodies (optimal dilutions are obtained via individual labeling).
c. Rinse the sections in TBS 5–10 minutes, twice.
d. Incubate the sections for 30 minutes at room temperature in mixture of two secondary antibodies (optimal dilutions are obtained via individual labeling).
e. Rinse the sections in TBS 5–10 minutes, twice.
f. Incubate the sections for 30 minutes at room temperature in the mixture of PAP and APAAP complex (optimal dilutions are obtained via individual labeling).
g. Rinse the sections in TBS 5–10 minutes, twice.
h. Display HRP: develop 2–5 minutes in DAB substrate solution.
i. Rinse the sections in TBS 5–10 minutes, twice.
j. Display AKP: develop 10–15 minutes in naphthol phosphate plus solid blue.
k. Rinse the sections in TBS 5–10 minutes, twice.
l. Place a drop of mounting medium and cover the sections with cover slips.

10.3.3.3 Notes
a. Paraffin blocks containing samples should be cut into 5–7 µm thick tissue sections.
b. TBS should be used as rinse buffer in the procedure of staining.
c. Endogenous AKP has no effect on the experimental results generally. If necessary inhibition (such as frozen sections) is needed, the following methods can be used:
   i. Levamisole can be added into AKP chromogenic medium to a final concentration of 1–10 mmol/L to inhibit endogenous AKP. AKP in APAAP complex, which originates from calf intestine, will not be affected because this operation has no influence on AKP of intestine. However, when the samples are made from intestine, this method can be replaced by the following methods:
   ii. Incubate the sections in 20% acetic acid for 20–30 minutes.
   iii. Incubate the sections in 2.5% periodic acid (or plus 0.3% H_2O_2) for 20–30 minutes.

d. Normal serum used as block comes from the same animal species with two kinds of secondary antibodies, or it can be omitted.
e. Eliminate cross-reaction: The second secondary antibody could combine with the first primary antibody because of the cross-reaction between different species Ig, that is, goat antirabbit IgG could react with mouse IgG, which will cause false double-staining results. Thus, it is necessary to perform antibody cross test, for example, incubate the sections in goat antirabbit IgG after mouse IgG or incubate the sections in goat antimouse IgG after rabbit IgG. Negative staining means no cross-reaction.

10.3.3.4 Evaluation
The great advantages are that there is no need of elution after the first staining and application of the mixtures of antibodies, but the control tests are necessary.

10.3.4 Immunohistochemistry double staining of primary antibodies from the same species

There are different kinds of enzymes applied in this method. The common method is immunoenzyme single labeling with different chromogenic mediums.

10.3.4.1 Basic methods

10.3.4.1.1 Color remaining method
Remove residual antibody and enzyme from sections after the first staining while one colored final reaction products are left at antigenic sites. Then perform the second staining by the same method as the first one, and display another antigen via another color caused by different electron donors. Using this method, two kinds of antigens are revealed on the same section. The appearance of blend color, such as color dusty purple, a combination of brown from DAB reaction and blue from CN reaction, indicates that two antigens exist in the same cell. The staining procedure is detailed as follows:
a. Treatment of sections, inhibition of endogenous enzyme and block by normal serum are previously described.
b. Finish the first staining by PAP immunoperoxidase, and perform color development by $CN-H_2O_2$.
c. Rinse the sections two times in TBS (5–10 minutes).
d. Antibody elution is performed by oxidation method.
e. Rinse the sections two times in TBS (5–10 minutes).
f. Finish the second staining by PAP immunoperoxidase, and perform color development by $DAB-H_2O_2$.
g. Rinse the sections two times in TBS (5–10 minutes).
h. Place a drop of mounting medium and cover with cover slips.

10.3.4.1.2 Color removal method

Take photos and remember view sites after completing the first staining. Use an organic solvent to dissolve and remove the final reaction products, and then rinse off antibody complex formed in the first staining. Continue the second staining and take photos in the same sites as before. Compare these two pictures to determine whether two antigens exist in the same location or not (Fig. 10.2). Staining procedures are detailed as follows:

a. Treatment of sections, inhibition of endogenous enzyme and block by normal serum are previously described.
b. Finish the first staining by PAP immunoperoxidase, and perform color development by CN-H_2O_2.
c. Rinse two times in TBS (5–10 minutes).
d. Mount sections with TBS temporarily, and take photos (add TBS from the coverslip edge at any time to prevent drying).
e. Take off coverslip, rinse in TBS, dispose of sections in alcohol-xylene-alcohol to remove CN blue reaction products and rinse in DDW.
f. Antibody elution is performed by oxidation method, rinse two times in TBS (5–10 minutes).
g. Finish the second staining by PAP immunoperoxidase, and perform color development by DAB-H_2O_2 or CN-H_2O_2.
h. Rinse two times in TBS (5–10 minutes/rinse) and mount sections.
i. Search for the accurate first photo site under microscope, and take another photo.

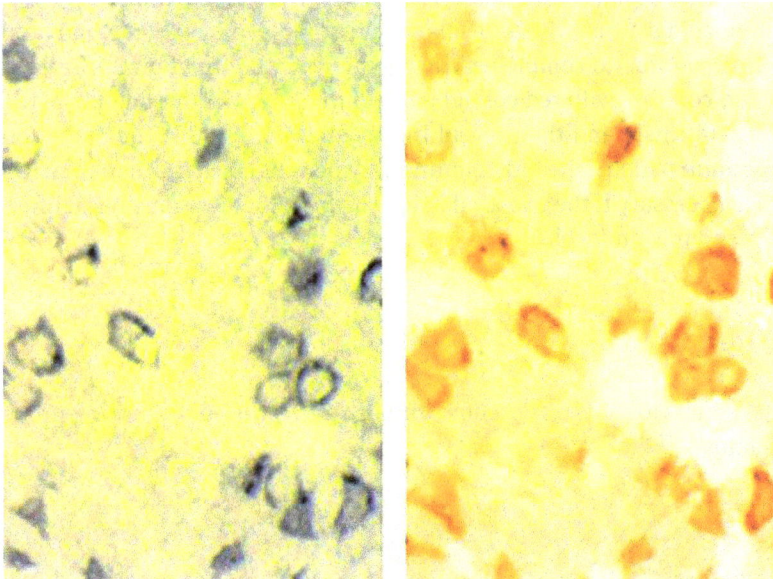

Fig. 10.2: The color removal method is introduced between GnRH-related peptides immunostaining (PAP method, left) and GnRH immunostaining (PAP method, right).

10.3.4.2 Notes

Antibody elution is the key step. A large number of rigorous controls must be performed to prove that antibodies are eluted completely and the second antigen is unaffected. Currently, it is very easy to obtain two kinds of primary antibodies originated from different animal species because of the appearance of various kinds of monoclonal antibodies. Therefore, the application of this method is reduced.

10.4 Immunoenzyme-immunofluorescence double staining

Immunoenzyme-immunofluorescence double-staining technology can be used to reveal two antigens on the same section. Display the first antigen by immunoenzyme at first, and then localize the second one by indirect immunofluorescence. After finishing staining, observe the results and take photos under fluorescence and bright light microscope simultaneously. Alternatively, record the immunopositive staining of the same view site via fluorescence and ordinary microscope, respectively, analyze the pictures and draw the conclusion. Also, elute the antibody complex of the first staining by oxidation, which may enhance fluorescence intensity but decrease the depth of color brown caused by DAB.

10.5 Immunoenzyme-immunogold double staining

The indication of two antigens on the same section is performed by immunoenzyme (PAP) plus immunogold staining (IGS). Display the first antigen by PAP method, and apply CN substrate solution to complete chromogenic reaction (final products show blue) to strengthen contrast with red colloid gold. Elute the antibody complex of the first staining by acid rinse, and then reveal the second one by IGS. The operation sequence is important; usually, the PAP immunoenzyme is finished first and then IGS follows. Elution is necessary if the two kinds of primary antibodies come from different animal species in this protocol.

Review question

What are the procedures to show nitric oxide synthase and GnRH in the same neurons in the hypothalamus?

11 Lectin histochemistry

The carbohydrate chains or oligosaccharide chains, which locate on the surface of cells, are important identification cell markers. All the changes in carbohydrate compositions on the surface of the cells or inside the cells are related to cell functions, cell behaviors, cell differentiations, cell maturation, cell malignant transformations, and so on. Generally, carbohydrate histochemistry method is used to identify polysaccharide, such as Periodic acid-Schiff (PAS) reaction. However, lectin histochemistry provides a simple and sensitive method for detecting glycosyl located inside the tissues or cells so that the changes of cell functions will be detected.

Lectins are a group of proteins or glycoproteins that are nonimmunogenic and can combine with carbohydrates. Thus, lectin can agglutinate cells and/or make carbohydrate combination precipitation. Lectin is not an antibody but can specifically combine with certain glycosyl. Most lectins are derived from plants and can make the red blood cell agglutinate. Now, hundreds of kinds of lectins have been isolated. The special merit of lectins is that they can sensitively and specifically identify and localize carbohydrates. Lectins can identify the tiny differences between carbohydrate chains so as to distinguish cell subpopulation and cell behavior related to the expression of carbohydrates.

11.1 Characteristics and application of lectin

11.1.1 Structure of carbohydrate in the tissue

The basic structure of monosaccharide is a six-carbon ring or chain. However, in the hundreds of monosaccharides in nature, only about seven kinds of monosaccharides are found in mammalian cells, namely, mannose, glucose, galactose, fucose, *N*-acetyl galactose, *N*-acetyl glucosamine, and sialic acid, and all of them are able to combine with the specific lectins.

11.1.2 Characteristics of lectins

11.1.2.1 Source and nomenclature of lectin
Lectins widely exist in nature, especially the seeds of legumes; hence, they are named according to their sources of plants or animals, such as peanut agglutinin (PNA), soybean agglutinin, wheat germ agglutinin (WGA), etc. Their abbreviations come from ordinary or Latin name, such as *Lens culinaris* lectin, *Ricinus communis* agglutinin, *Ulex europaeus* lectin, etc.

DOI 10.1515/9783110531398-011

11.1.2.2 Structural characteristics of lectin

Most lectins are glycoproteins, but there are at least three kinds of commonly used agglutinin, such as conconvalina (Con A), PNA, and WGA, which are proteins and do not combine with carbohydrate chain themselves. All lectin molecules are composed of two subunits polymerization at least, and some can be up to 18 subunits. Each subunit has one carbohydrate-binding site.

11.1.2.3 Combination features of lectins

The exact physiological function of lectin is unknown. However, lectins have been widely used in biomedical field because they have high specificity in combination with glycosyl. The features of binding between lectins and carbohydrate are as follows:

a. The specificity of sugar-binding aspect of lectin can be designated by the soluble monosaccharide that can inhibit the lectin combination, but actually its specificity of monosaccharide binding may be different from the specific glycosyl in tissue. In other words, it is oligosaccharide chains that bind to lectins but not monosaccharide. Thus, monosaccharide inhibits binding between oligosaccharide chains and lectins (Fig. 11.1). The oligosaccharide specificity of most lectins is unclear.

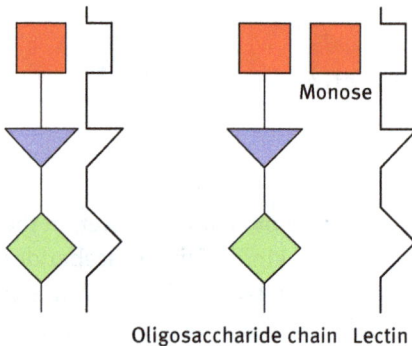

Monose

Oligosaccharide chain   Lectin

Fig. 11.1: Combination between lectins and oligosaccharide chains.

b. The presence of heavy metal ions such as Ca^{2+}, Mn^{2+}, and Mg^{2+} is important to maintain the activity of the lectin-binding site.
c. The binding features of lectin depend on the pH level.
d. All lectins have two carbohydrate-binding sites at least.
e. The binding of lectins and carbohydrate is reversible.

11.1.3 Application of lectins

Although any lectin is not cell specific or tumor specific, lectin, acting as a useful tool, can identify certain cell type, cell differentiation, and maturation. In other words, lectins can indicate the related messages about tumor, such as tissue source, degree of differentiation, recurrence risk, and so on.

11.2 Application of lectin histochemistry

Lectins can be labeled by fluorescent substances, enzyme, biotin, colloidal gold, etc. Therefore, lectins can be used in histochemical or immunohistochemical study.

11.2.1 Principle

11.2.1.1 Basic mechanism
Lectins combine with tissue or cell specifically. A certain method should be used to make the combination to be visible under light microscopy and electron microscopy. The commonly used lectin in histochemistry includes direct method, indirect method, and carbohydrate-lectin-carbohydrate sandwich method.

11.2.1.2 Condition

11.2.1.2.1 Tissue preparation
Fresh and untreated tissues are used to prepare smear or frozen section; therefore, carbohydrate was not sealed by fixative. Lectin histochemistry can show the carbohydrate ideally.

11.2.1.2.2 Buffer
The buffer used for diluting lectin and section rinse is Tris-buffered saline (pH 7.4). A small amount of a variety of two valence metal ions should be added to the buffer to keep the activity of the lectin-binding site. The ingredients of buffer used in lectin histochemistry are shown as follows:

Tris	6.08 g
NaCl	8.7 g
Double-distilled water	Make up to 100 mL, adjust pH level to 7.4
1% $CaCl_2$	1.5 mL
1% $MnCl_2$	2 mL
1% $MgCl_2$	2 mL

11.2.1.2.3 Exposing glycosyl on tissue slice
Proteolytic enzyme is available to digest and expose the sealed glycosyl. The detection rate of glycosyl is increased.

11.2.1.2.4 Add protein and detergents

Some proteins such as 0.1%–10% bovine serum albumin that does not contain glycosyl can stabilize lectins and weaken the nonspecific staining.

11.2.2 Direct method

11.2.2.1 Theory

Lectins are labeled with fluorescein isothiocyanate (FITC), Horseradish peroxidase (HRP), and colloidal gold. The labeled lectin incubates tissue section directly. Finally, the section can be observed directly with fluorescent microscopy. The enzyme is shown with a color-developing agent. Then the section can be observed with normal light microscopy (Fig. 11.2).

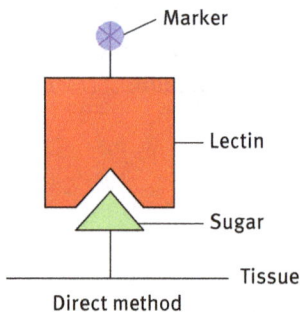

Fig. 11.2: Direct method.

11.2.2.2 Staining procedures

a. Frozen sections are dried in air (with or without acetone or methanol fixation); paraffin sections are dewaxed to water.
b. Rinse the sections in buffer.
c. Drop the labeled lectins on the sections, and incubate for 30 minutes at room temperature.
d. Rinse the sections three times with buffer, 2 minutes each.
e. Seal the sections directly or after dehydration and clearance.

11.2.2.3 Valuation

a. The direct method is the simplest method. It is especially suitable for frozen section, tissue culture, cell smear, and other untreated tissue preparation.
b. This method possesses strong specificity and higher success rate.
c. This method chooses the optimal dilution degree of lectin and incubation time.
d. FITC-labeled lectin is more reliable to show glycosyl on frozen section. The endogenous enzyme of specimen must be sealed if the enzyme-labeled lectin is applied.
e. The sensitivity of the paraffin section is not so good that high concentrations of lectin or longer reaction time may be applied.

11.2.3 Antibody method

11.2.3.1 Theory
The section is incubated in the lectin solution to make lectin bind to carbohydrate chain. Then the antibody of lectin is added after rinse (Fig. 11.3A). The special primary antibody can be labeled with FITC, HRP, and colloidal gold or not. If the unlabeled primary antibody was selected, the labeled secondary antibody should be used (Fig. 11.3B). If the unlabeled secondary antibody is selected, the PAP complex will be used (Fig. 11.3C).

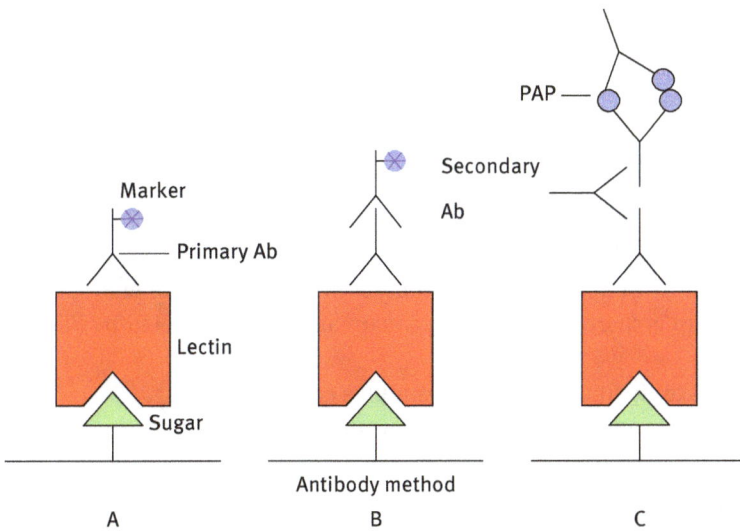

Fig. 11.3: Antibody method.

11.2.3.2 Staining procedures

11.2.3.2.1 Method for labeled primary antibody (one-step method)
a. Paraffin sections are dewaxed to water.
b. Immerse the sections in 3% H_2O_2 for 10 minutes. The endogenous enzymes are inhibited. This step is disregarded if the primary antibody is labeled by FITC or colloidal gold.
c. Rinse the sections three times in buffer, 2 minutes each.
d. Immerse the sections in unlabeled lectin (10 µg/mL) for 30 minutes at room temperature.
e. Rinse the sections three times in buffer, 2 minutes each.
f. Immerse the sections in labeled primary antibody (1:100) for 30 minutes.
g. Rinse the sections three times in buffer, 2 minutes each.
h. Observe microscope after DAB-H_2O_2 color development.

11.2.3.2.2 Method for labeled secondary antibody (two-step method)
a. Paraffin sections are dewaxed to water.
b. Immerse the sections in 3% H_2O_2 for 10 minutes. The endogenous enzymes are inhibited. This step is disregarded if the primary antibody is labeled by FITC or colloidal gold.
c. Rinse the sections three times in buffer, 2 minutes each.
d. Immerse the sections in unlabeled lectin (10 µg/mL) for 30 minutes at room temperature.
e. Rinse the sections three times in buffer, 2 minutes each.
f. Immerse the sections in unlabeled primary antibody (1:100) for 30 minutes.
g. Rinse the sections three times in buffer, 2 minutes each.
h. Immerse the sections in labeled secondary antibody, 1:50–1:100, 30 minutes.
i. Rinse the sections three times in buffer, 2 minutes each.
j. Observe microscope after DAB-H_2O_2 color development.

11.2.3.2.3 PAP method
a. Paraffin sections are dewaxed to water.
b. Immerse the sections in 3% H_2O_2 for 10 minutes. The endogenous enzymes are inhibited. This step is disregarded if the primary antibody is labeled by FITC or colloidal gold.
c. Rinse the sections three times in buffer, 2 minutes each.
d. Immerse the sections in unlabeled lectin (10 µg/mL) for 30 minutes at room temperature.
e. Rinse the sections three times in buffer, 2 minutes each.
f. Immerse the sections in unlabeled primary antibody (1:100) for 30 minutes.
g. Rinse the sections three times in buffer, 2 minutes each.
h. Immerse the sections in unlabeled secondary antibody, 1:50–1:100, 30 minutes.
i. Rinse the sections three times in buffer, 2 minutes each.
j. Immerse the sections in PAP complex, 1:50–1:100, 30 minutes.
k. Rinse the sections three times in buffer, 2 minutes each.
l. Immerse the sections in DAB-H_2O_2 for color development, and then undergo dehydration, clearing, and sealing.

11.2.3.3 Evaluation
a. The primary antibody should be purified IgG. Thus, dyeing is clear and the background dyeing is low.
b. To reduce the background staining, sections can be treated with bovine serum albumin, or bovine serum albumin is added in the preparation of lectins buffer.

11.2.4 Biotin-labeled method

11.2.4.1 Principle

First, lectin is labeled with biotin. The labeled lectin binds to the carbohydrate chain on the cell surface. Second, labeled avidin is added into the reaction system (Fig. 11.4A), or the ABC complex is added into the reaction system (Fig. 11.4B). Thus, at least the label is visualized to show the combination site of lectin.

11.2.4.2 Staining procedures

a. Dewax paraffin sections to water.
b. Immerse the sections in 0.1% trypsin (containing 0.1% $CaCl_2$) for 10–30 minutes at 37°C.
c. Rinse the sections three times in buffer, 2 minutes each.

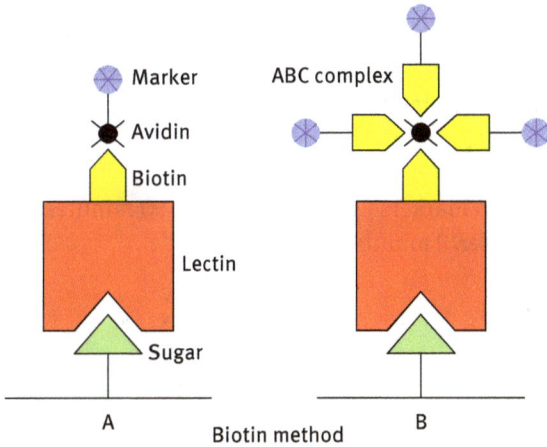

Fig. 11.4: Biotin-labeled method.

d. Immerse the sections in 3% H_2O_2 for 10 minutes. The endogenous enzymes are inhibited.
e. Rinse the sections three times in buffer, 2 minutes each.
f. Immerse the sections in biotin-labeled lectin (2–10 µg/mL) for 30–60 minutes.
g. Rinse the sections three times in buffer, 2 minutes each.
h. Immerse the sections in labeled avidin (1–10 µg/m) for 30 minutes, or ABC complex for 30–60 minutes.
i. Rinse the sections three times in buffer, 2 minutes each.
j. Observe under fluorescence microscope (fluorescent labeling) under light microscopy (enzyme labeling or ABC compound visualized by DAB-H_2O_2).

11.2.4.3 Valuation

Biotin-labeled method (especially the application of ABC complex) is a relatively simple and sensitive method to show carbohydrate chain. Moreover, nonspecific staining of the background is very low.

11.2.5 Carbohydrate-lectin-carbohydrate sandwich method

11.2.5.1 Principle

Excess lectin is added on sections to make sure that the lectin-binding sites are incompletely dominated by glycosyl in tissue. Then glycosylation markers are added to combine with the lectin. Finally, markers are displayed (Fig. 11.5).

11.2.5.2 Staining procedures

a. Dewax paraffin sections to water.
b. Immerse the sections in 3% H_2O_2 for 10 minutes. The endogenous enzymes are inhibited.
c. Rinse the sections three times in buffer, 2 minutes each.
d. Immerse the sections in lectin (10–20 µg/mL) for 30 minutes.
e. Rinse the sections three times in buffer, 2 minutes each.
f. Immerse the sections in glycosylation markers (20 µg/mL) for 30–60 minutes.
g. Rinse the sections three times in buffer, 2 minutes each.
    Display the marker.

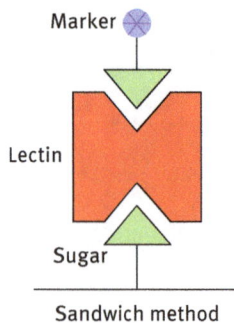

Fig. 11.5: Carbohydrate-lectin-carbohydrate sandwich method.

11.2.5.3 Valuation

a. Because of using unlabeled lectin, degeneration may not occur. Unlike the antibody method, it only involves the combination of carbohydrate. Thus, sensitivity and specificity are higher than that of the antibody method.
b. The widely used glycosylation enzyme is HRP. HRP has a large number of mannose bases, which make it applied in exploring lectin that own the characteristics of mannose combination.

11.2.6 Control test

11.2.6.1 Positive control
Positive control is used to prove that the lectin, all dyeing reagents, and methods are effective.
a. Lectin agglutination control test: Most lectins can combine with red blood cells. This feature can be used as a positive control.
b. Tissue control test: Red blood cells and vascular endothelial cells on the sections bind to lectins, which can be used as positive control.

11.2.6.2 Negative control
Checking the specificity of the dyeing and monosaccharide specificity is the purpose of negative control.
a. The binding sites in the tissue are sealed with excessive unmarked lectin before the labeled lectin is introduced in the direct method.
b. Lectin is not used during the dyeing process of the indirect method.
c. Add unlabeled glycoprotein between steps 4 and 6 during the carbohydrate-lectin-carbohydrate sandwich method.
d. Lectin or labeled lectin is diluted in 0.1–0.2 mol/L inhibitory monosaccharide solution, incubated for 30 minutes, and then used for dyeing.

Knowledge links: History of lectin application

In 1919, J. Sumner purified Con A, and in 1936, he found that Con A could precipitate glycogen. The lectin specificity for blood type antigens was found in 1940, and in 1952, W. Watkins and W. Morgan demonstrated that blood type antigens are saccharides by lectin and glycoside hydrolase, and lectins were widely used to replace hemagglutinin in 1954. It was found that some lectins could stimulate the resting lymphocyte to divide in 1960, and during 1960–1980, the function of agglutination of lectins for some animal tumor cells was discovered. Nowadays, the *Galanthus nivalis* agglutinin gene, the peas exogenous agglutinin gene, the wheat germ agglutinin gene, the amaranth lectins gene, and the ribosome-inactivating protein gene have been used in gene engineering against plant disease resistance, and lectins have been introduced to diagnose malignant tumor.

12 Progresses of *in situ* display

The newly created methods of demonstration of chemical complexes mainly embody the progresses of *in situ* display.

12.1 Envision method

The envision method belongs to the indirect method. Unlike the common indirect method in which the secondary antibodies are labeled by enzymes, in the envision method, the enzyme molecules are connected by a long inert polymer, such as glucosan, forming an enzyme-glucosan complex, and then the complex is labeled by secondary antibodies to create a secondary antibody-glucosan-enzyme complex (Fig. 12.1). In the staining procedures, the primary antibodies are used to detect the antigens in tissue sections, and then the complexes are added to combine with the primary antibodies. Because the complex contains large amounts of enzymes, the final products of enzymatic reaction are greatly magnified, which means that the envision method is very sensitive. Because glucosan is exogenous, nonspecific staining can be disregarded.

If the glucosan-enzyme complexes are directly labeled by primary antibodies and then are used to detect the antigens in tissue sections, it is known as the enhanced polymer one-step staining method.

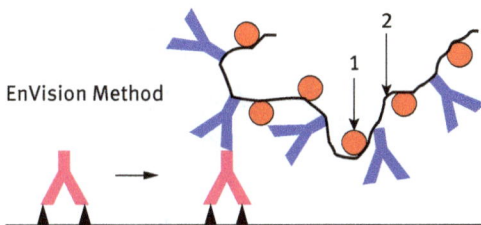

Fig. 12.1: The envision method. Arrow 1: enzyme; arrow 2: glucosan.

12.2 Catalyzed signal amplification method

The catalyzed signal amplification (CSA) method is a modified streptavidin-perosidase (SP) method. In the CSA method, after the primary antibodies are used to detect the antigens in tissue sections, the biotin-labeled secondary antibodies, the streptavidin (SA)-Horseradish peroxidase (HRP), and the biotin molecule-tyramine group are introduced one after another. The HRP catalyzes the biotin molecule-tyramine group, forming large amounts of biotin deposits around the antigens. When SA-HRP complexes are added to tissue sections again after rinse, the large amount of HRP will be aggregated *in situ*, which results in the significant final signal magnification (Fig. 12.2).

DOI 10.1515/9783110531398-012

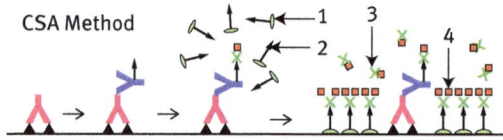

Fig. 12.2: The CSA method. Arrow 1: tyramine group; arrow 2: biotin; arrow 3: SA; arrow 4: HRP.

12.3 Ferric oxide alternative method for HRP

The purpose of this method is to save the expensive HRP for cheaper laboratory expenses. In this method, the activities of H_2O_2 catalysis of nano-γ-ferric oxide (Fe_2O_3) particles are greatly enhanced by a Prussian blue coat and modified by a bovine serum albumin coat. Protein A connects the complex particles and the Fc segments of antibodies. The final structures will catalyze the tetramethylbenzidine (TMB) as substrates (Fig. 12.3).

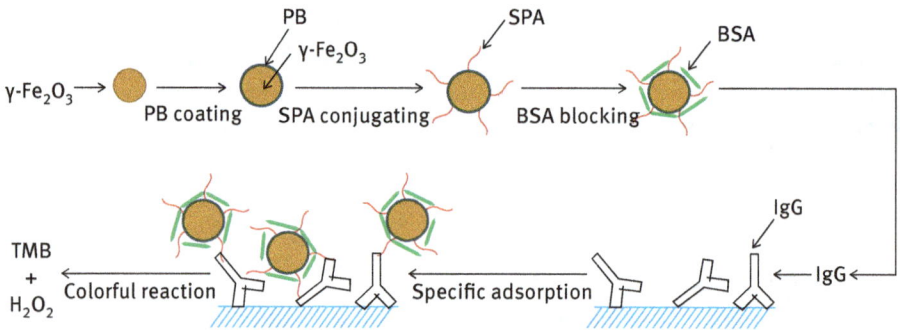

Fig. 12.3: Ferric oxide used as HRP.

Knowledge links (http://www.ncbi.nlm.nih.gov/ pubmed/21466356)

Numerous immunohistochemical stains have been shown to exhibit exclusive or preferential positivity in breast myoepithelial cells relative to their luminal/epithelial counterparts. These myoepithelial markers provide invaluable assistance in accurately classifying breast proliferations, especially in core biopsies. Although numerous myoepithelial markers are available, they differ in their sensitivity, specificity, and ease of interpretation, which may be attributed, to a large extent, to the variable immunoreactivity of these markers in stromal cells, including myofibroblasts, vessels, luminal/epithelial cells, and tumor cells.

13 *In situ* hybridization histochemistry technology

In situ hybridization histochemistry (ISHH) is a powerful technique that combines molecular biology with morphological science. ISHH is a type of hybridization that uses a labeled complementary DNA or RNA strand (i.e., probe) to localize a specific DNA or RNA sequence in cells or tissue sections. Then the hybridization signal is detected by histochemistry or immunohistochemistry methods and examined with light microscope or electron microscope.

13.1 Basic theory

Two complementary mononucleotide strands interact with each other to form stable hybrids by hydrogen bonds between the two complementary base pairs at suitable temperature and salt concentration. Thus, a nucleotide strand with known base sequence is labeled with either radioisotopic tracer or nonradioactive marker to generate a probe. Then this probe hybridizes with the target nucleotide in the tissue section, cell smear, cultured cells or chromosome sample. The labeled probes can be localized and quantified at light microscope or electron microscope levels (Fig. 13.1). This technique allows the localization of specific nucleotide sequences as small as 10 to 20 copies of mRNA or DNA per cell.

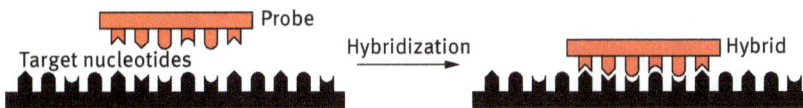

Fig. 13.1: Schematic diagram of the basic theory of ISHH.

ISHH is classified into DNA-DNA hybridization, DNA-RNA hybridization and RNA-RNA hybridization according to the nature of probe and target nucleic acid. According to the nature of marker, ISHH is classified into radioactive and nonradioactive ISHH. ISHH is classified into direct or indirect method on the basis of direct or indirect detective technique.

13.2 Probe preparation

Probes are marker-labeled nucleotide molecules that are complementary with the target nucleotide. They should be highly specific to the target nucleotide with small molecules, strong tissue penetrating power and high specific radioactivity. The probe

DOI 10.1515/9783110531398-013

preparation should be convenient, and high resolving power of the detection of labeled probes is necessary.

13.2.1 Category of probes

Three types of probes are frequently used in the *in situ* hybridization. They are DNA probe, cRNA probe and oligonucleotide probe.

13.2.1.1 DNA probe
Double-stranded recombinate DNA (cDNA) probes are most frequently used at present. These probes have constant length, high sensitivity and good labeling effect. The major defect is that cDNA probes are double stranded.

13.2.1.2 cRNA probe
Reverse transcriptase polymerase chain reaction (RT-PCR) is performed for the amplification of the target gene, and the PCR fragments are cloned into the polylinker site of a transcription vector, which contains a promoter for SP6 and T7 RNA polymerase. The recombinant plasmids serve as templates for the generation of the PCR products of the target gene with the SP6 and T7 primers. The recovered PCR products are used to synthesize digoxigenin (DIG)-labeled antisense and sense RNA probes by in vitro transcription with SP6 and T7 RNA polymerases and RNA labeling mix that contains ATP, CTP, GTP, UTP and marker-UTP. The process of cRNA probe preparation is actually complex. RNA probes are easily degraded by RNase.

13.2.1.3 Oligonucleotide probe
Oligonucleotide probes are artificially generated by DNA synthesizer according to the already-known nucleotide sequence of the target gene.

13.2.2 Principal and method of probe labeling

Probes can be labeled by radioactive isotope or nonradioactive marker. The procedure of isotope labeling incorporates ^{3}H-, ^{35}S- or ^{32}P-labeled nucleotides or deoxynucleotides in the probe base sequence or links them at the 3′ or 5′ terminal of probes. The radioactive isotope method is highly sensitive, but the defects include complex operation, long exposure time, high background and radioactive injury. In addition, this method is affected by half-life of isotope.

Nonradioactive labeling method has such good qualities as safety, stability, good cellular localization, quick coloration and absent radioactive pollution.

Biotin and DIG are often used as markers. For biotin-labeled probes, the target nucleic acids are visualized after hybridization with enzyme-labeled antibiotin antibody or avidin-biotin enzyme. DIG is a steroid half-antigen with small molecular mass, and it is not affected by endogenous similar substances. Fluorescence can also be used to label probes, and it is observed with fluorescence microscope after hybridization.

13.3 Procedure of ISHH

13.3.1 Basic procedure

The basic flow diagrams of ISHH are similar, although the methods vary because of different tissues, target nucleotides and probe types (Fig. 13.2).

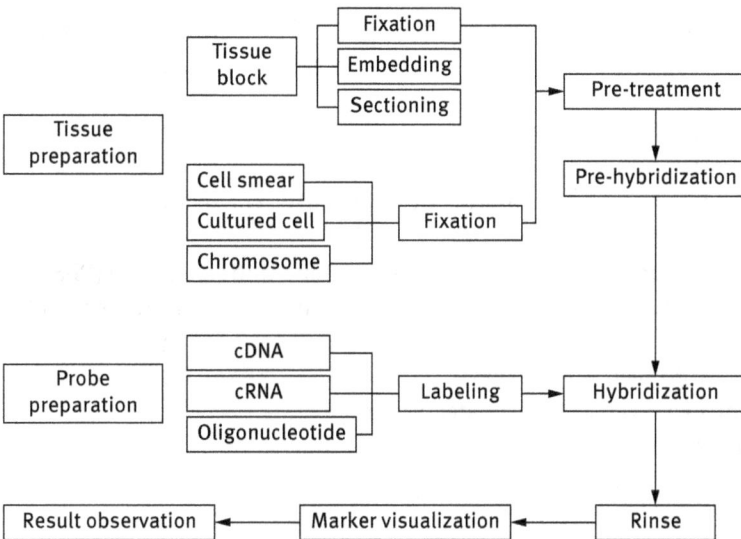

Fig. 13.2: Basic procedures of ISHH.

13.3.2 Detection of mRNA on paraffin-embedded sections with DNA probes

13.3.2.1 Tissue preparation
Basic theories
a. Preserve fine histological and cytological structures.
b. Preserve DNA or RNA in cells.

  c. Prevent RNase pollution during detection of mRNA. The methods are as follows:

      i. Wear gloves and mouth muffle in the experiments because hands and saliva contain RNase.

      ii. After routine rinse, the glassware should be soaked into RNase inhibitor such as 0.1% diethyl pyrocarbonate (DEPC)-treated H_2O for 2 hours at 37°C. Rinse the glassware several times with sterilized distilled water and autoclave to remove DEPC. Bake all glassware for 4 hours at 180°C.

      iii. Assume that disposable plastic ware is RNase free. Eppendorf tubes and tips are new, and autoclave them before use.

      iv. Fire the scissors and pliers with flame.

      v. Prepare all solutions and buffers with DEPC-treated water. DEPC (0.05%–0.1% final concentrations) is added to all solutions. Allow the solutions to stand at room temperature overnight or stir on a magnetic stirrer for 20 minutes at room temperature, and autoclave (1.034×10^5 Pa, 15–30 minutes) to remove DEPC.

  d. Prevent the shedding of tissue slices from the glass slides during hybridization and rinse: Apply sticking reagents and rinse the sections thoroughly.

Fixation, embedding and sectioning

  a. Obtain fresh tissues with RNase-free apparatus.

  b. Fix the tissues in ice-cold 4% paraformaldehyde at 4°C. The fixation should less than 24 hours to prevent loss of mRNA.

  c. Rinse the tissues: 30 minutes, twice, with 1× phosphate-buffered saline (PBS).

  d. Perform routine dehydration, clearing and embedding. Frozen sections are also used.

  e. The tissue is cut into sections (4–5 μm)

  f. Gain the sections with 0.1% DEPC-treated water.

  g. Bake the sections at 37°C for 1–3 days.

  h. Store the sections at –20°C.

Section treatment

  a. Immerse the sections in 1 mol/L HCl for 24 hours.

  b. Immerse the sections in 95% ethanol for more than 2 hours.

  c. Rinse the sections with water.

  d. Rinse the sections three times with distilled water. Dry the sections.

  e. Bake the sections for 4 hours at 180°C to remove nuclease.

  f. Immerse the sections in gelatinum-chromium potassium sulfate solution (or polylysine solution).

  g. Dry the sections thoroughly. Store the sections under dust and RNase-free conditions. Wear sterilized gloves during these steps.

Solutions preparation

a. 10× PBS stock solution

NaCl	40.0 g
KCl	1.0 g
KH_2PO_4	1.2 g
Na_2HPO_4	7.2 g
Double distilled water (DDW)	400 mL, adjust pH to 7.4 with NaOH
DDW	Make up to 500 mL. Autoclave and store at room temperature.

b. Gelatinum-chromium potassium sulfate sticking reagent

Gelatinum	2.5 g
0.1% DEPC	500 mL, heat to 70°C to dissolve gelatinum. After cooling, add 2.5 g $CrK(SO_4)_2$, and stir into solution.

c. 0.1% poly-L-lysine sticking reagent

Poly-L-lysine	100 mg
DDW	100 mL

d. 1.4.4 0.1% (v/v) DEPC-treated H_2O

DEPC	500 μL
DDW	500 mL

Allow the solution to stand at room temperature overnight or stir the solution on a magnetic stirrer for 20 minutes at room temperature and autoclave to remove DEPC.

13.3.2.2 Prehybridization pretreatment

13.3.2.2.1 Objectives and basic principles

a. Allow the maximum amount of probe to penetrate the tissue and elevate hybridization sensitivity. HCl, protease or detergent such as Triton X-100 are often used.
b. Acetic anhydride is used to reduce nonspecific background by promoting basic groups acetylation in tissue and neutralizing electric charges.
c. If *in situ* hybridization is used to detect DNA in nucleus, RNase needs to be used to digest RNA in tissue to increase the hybridization specificity.

13.3.2.2.2 Procedures

The following methods are often used for the first two objectives mentioned previously:

a. Perform routine deparaffin and rehydration.
b. Treat the sections with 0.2 mol/L HCl for 20 minutes at room temperature. Rinse the sections with distilled water for 2 minutes.
c. Treat the sections with protease K (1 μg/mL) for 30 minutes at 37°C. Rinse the sections with distilled water for 2 minutes.

d. Postfix the sections with 4% paraformaldehyde for 4 minutes at room temperature.
e. Rinse the sections for 2 minutes, twice, with 1× PBS, then with distilled water for 2 minutes.
f. Incubate the sections in freshly mixed 0.25% (v/v) acetic anhydride for 15 minutes at room temperature. Rinse the sections with distilled water for 2 minutes.
g. Dehydrate the sections in a graded series of ethanol solutions (increasing concentrations of ethanol) and air-dry.
h. Perform prehybridization and hybridization. The sections can be stored at 4°C.

For frozen sections, change procedures c–e as follows:
a. Treat the sections with 2× saline sodium citrate (SSC) for 30 minutes at 70°C. Rinse the sections with distilled water for 2 minutes.
b. Use pronase (0.25 mg/mL) at room temperature for 10 minutes.
c. Use hydroxyl group aminoacetic acid (2 mg/mL) at room temperature for 5 minutes. Rinse the sections for 2 minutes, twice, with 1× PBS.

Announcements
a. Pretreatments are not necessary, and they are related to fixatives and probe size. Tissues fixed with nonaldehyde fixatives such as ethanol, acetic acid, etc., do not need pretreatments. When probes less than 300 bp hybridize with the target nucleotide in tissues fixed with paraformaldehyde, pretreatments are unnecessary.
b. Tissue morphology and structure should not be damaged in the process of pretreatments.

The following methods are used for the third objective mentioned previously (digestion of RNA):
a. Perform routine deparaffin and rehydration.
b. Treat with 10 mmol/L Tris-HCl, 2 minutes.
c. RNase incubation buffer (100 µg/mL), 37°C, 30 minutes.
d. Treat with 10 mmol/L Tris-HCl, 2 minutes.
e. Rinse with distilled water for 2 minutes.
f. Dehydrate the sections in a graded series of ethanol solutions (increasing concentrations of ethanol) and air drying.
g. Perform prehybridization and hybridization. The sections also can be stored at 4°C.
h. The above-mentioned treatments can serve as negative control in mRNA detection.

13.3.2.2.3 Solution preparation
a. Protease K stock solution (1.5%)

Protease K	15 mg
100 mmol/L Tris-HCl (pH 8.0)	1 mL

b. Protease K working solution (1 μg/mL)

100 mmol/L CaCl$_2$	0.6 mL
1 mol/L Tris-HCl (pH 8.0)	0.3 mL
DDW	Make up to 30.0 mL
1.5% protease K	2 μL

c. Pronase working solution

Pronase	12.5 mg (0.25 mg/mL)
1 mol/L Tris-HCl (pH 7.5)	2.5 mL (50 mmol/L)
0.5 mol/L ethylenediaminetetraacetic acid (EDTA)	0.5 mL (5 mmol/L)
DDW	Make up to 50 mL

d. Acetic anhydride (use when fresh)

0.1 mol/L trolamine (pH 8.0)	50 mL
Acetic anhydride	125 μL

e. RNase incubation buffer

1 mol/L Tris-HCl (pH 7.5)	0.3 mL
3 mol/L NaCl	3.0 mL
DDW	Make up to 30.0 mL
RNase A (30 mg/mL)	100 μL (final concentration of 100 μg/mL)
RNase T$_1$ (3000 U/mL)	100 μL (final concentration of 10 U/mL)

f. 10 mmol/L Tris-HCl

1 mol/L Tris-HCl (pH 7.5)	0.3 mL
3 mol/L NaCl	3.0 mL
DDW	Make up to 30.0 mL

13.3.2.3 Prehybridization, hybridization and posthybridization treatment

13.3.2.3.1 Prehybridization

Objective: To saturate nonspecific binding sites of probe in tissues (such as protein, nucleic acid, polysaccharides, etc.) to reduce nonspecific background.

Procedures

a. Make fresh prehybridization buffer (vide infra).
b. Perform routine deparaffin and rehydration. (Omit b if pretreatment is already performed. If the sections are taken from –20°C, the sections should be dried at 37°C for 1 day to prevent tissue slices shedding from the glass slides.
c. Place the sections on a baking plate. Add 20 μL prehybridization buffer onto the tissue.
d. Incubate the sections in a humid chamber for 2 hours at 40°C.

13.3.2.3.2 Hybridization

Procedure

a. Remove prehybridization buffer. Rinse the sections with distilled water for 1 minute.

b. Dehydrate the sections in 100% ethanol for 1 minute and air-dry.
c. Encircle the tissue with a wax pencil.
d. Make hybridization buffer: Mix DNA probe with hybridization stock solutions E and F at a final probe-specific radioactivity of 1×10^6 cpm each section. Place the sections on a hot plate at 95°C for 10 minutes to unwind the DNA. Cool down immediately on ice. Hybridization stock solutions A, B, C and D are added. Mix the solutions.
e. Pipette 20 μL hybridization buffer onto the section.
f. Cover with RNase-free coverslip.
g. Use the rubber cement to seal the coverslip.
h. Incubate the sections in a humid chamber for 16 hours at 40°C.

Main points
a. Unwinding (usually heat denaturation) is necessary before hybridization if DNA probes are used.
b. If the DNA sequence of interest such as virus DNA is to be detected, add the DNA probes onto the section, then perform denaturation.
c. Probe concentration is very important to hybridization efficiency. Probe-specific radioactivity should be more than 0.5×10^5 cmp/μL hybridization buffer.
d. Salt ion concentration is also important to hybridization. The concentration of NaCl is usually 0.3–0.9 mol/L.
e. Perform pretesting to determine hybridization temperature (37°C–40°C) in 50% formamide solution.
f. The concentration of deionized formamide may affect hybridization specificity; 50% deionized formamide is recommended.

13.3.2.3.3 Posthybridization treatments
Objective: These treatments are necessary to remove redundant and nonspecifically bound probes; thus, they are helpful in reducing diffuse background staining.

Rinse with a high concentration of salt can decrease static electricity binding of probes and tissues. At least a stringency rinse (low concentration of salt, high temperature and containing formamide) is necessary. Do not make the sections dry during the process of rinse.
a. Remove the coverslip with a needle. Briefly rinse the sections with eluant A.
b. Rinse with eluant A for 1 hour at 50°C.
c. Rinse with eluant B for 1 hour at 37°C.
d. Rinse with eluant C, overnight, at room temperature, and stir lightly on a magnetic stirrer.
e. Dehydrate the sections in a graded series of ethanol solutions (increasing concentrations of ethanol). Perform radioautography or the sections are stored at 4°C.

Main points
a. Lower concentration of ions, higher temperature, longer time of rinse and more efficient effects can be reached.
b. Prevent the tissue slices shedding from the glass slides, especially in case of high temperature.
c. The effect of hybridization can be estimated by detecting the sections after rinse with a Geiger counter (except ^{3}H).
d. If RNA probes are used, treat the sections with RNase (10 µg/mL, 37°C, 30 minutes) after hybridization to degrade the single strand of RNA, which does not form hybrids. RNase does not degrade RNA-RNA hybrids. Posthybridization enzyme digestion to remove the free probes is unnecessary if cDNA probes are used.

13.3.2.3.4 Solution preparation (Tab. 13.1)

Tab. 13.1: Prehybridization buffer and hybridization buffer.

Component	Prehybridization buffer	Hybridization buffer
Hybridization stock solution A dextran sulfate	4.0 µl	4.0 µl
(added to stock solution A before use)	2.0 mg	2.0 mg
Hybridization stock solution B	0.5 µl	0.5 µl
Hybridization stock solution C	2.0 µl	2.0 µl
Hybridization stock solution D	0.5 µl	0.5 µl
Hybridization stock solution E	1.0 µl	1.0 µl
Hybridization stock solution F	10.0 µl	10.0 µl
DNA probe	—	2.0 µl
		(5×10^5 pm/µl)
Distilled H_2O	2.0 µl	—
Total volume	20.0 µl	20.0 µl

a. Hybridization stock solution A

Na_2HPO_4	0.352 g (0.25 mol/L)
NaCl	1.736 g (3 mol/L)
EDTA (0.5 mol/L)	500 µL (25 mmol/L)
Distilled water	Make up to 10 mL

b. Hybridization stock solution B (0.4 mol/L or 6.17% dithiothreitol [DTT] solution)

DTT	61.7 mg
Distilled water	Make up to 1.0 mL

Note: This solution is only needed in case of ^{35}S-labeled probes. Equal volume of distilled water is used instead of this solution in case of non ^{35}S-labeled probes.

c. Hybridization stock solution C (100× Denhardt's solution)

Bovine serum albumin (BSA)	2.0 g
Ficoll	2.0 g
Polyvinylpyrrolidone	2.0 g
Distilled water	Make up to 100 mL

d. Hybridization stock solution D (2% sodium pyrophosphate solution)

Sodium pyrophosphate ($Na_4P_2O_7$)	2.0 g
Distilled water	Make up to 10 mL

e. Hybridization stock solution E (1% nucleotide solution)

Sheared salmon sperm DNA	10 mg
Escherichia coli tRNA	10 mg
Distilled water	Make up to 1 mL

f. Hybridization stock solution F

Deionized formamide solution: 100 mL formamide + ion-exchange resin (such as Biorad AG501-X8, 20–50 mesh), stir for 30 minutes at room temperature and filter. Repeat twice. Dispense into aliquots and store at –20°C.

g. Eluants (Tab. 13.2)

Tab. 13.2: Chemical composition of eluants.

Component	Eluant A (1× SSC)	Eluant B (0.5× SSC)	Eluant C (0.05× SSC)
20× SSC	40 ml	10 ml	6.25 ml
10% Sodium pyrophosphate	4 ml	2 ml	12.5 ml
20% Sodium lauryl sulfate (SDS)	8 ml	–	–
Dithiothreitol (DTT)*	61.6 mg	30.8 mg	192.5 mg
50% Sodium thiosulfate*	16 ml	8 ml	50 ml
DDW make up to	800 ml	400 ml	2500 ml

*[35]S-labeled probe use only

h. 20× SSC

NaCl	175.3 g
Citrate sodium	88.2 g
DDW	Make up to 1000 mL

Adjust the pH to 7.04 with 1 mol/L HCl. Dispense into aliquots and sterilize by autoclaving.

Note: Triton X-100 can also serve as eluant.
1× SSC + 0.3% Triton X-100 at 42°C for 30 minutes.
0.5× SSC + 0.3% Triton X-100 at 42°C for 30 minutes, twice.
0.2× SSC + 0.3% Triton X-100 at 42°C for 30 minutes, twice.

13.3.2.4 Visualization

13.3.2.4.1 Autoradiography: used for detection of radioactive probes
Procedures
a. Place the nuclear emulsion in water bath for 20 minutes at 45°C.
b. Place the melted nuclear emulsion in stand-style slide jar at 45°C water bath.
c. Place the sections in the nuclear emulsion twice. Knock the sections to remove superfluous emulsion.
d. Place the sections in the box away from light. Dry the sections slowly for 4 hours (place wet papers in the box).
e. Place the sections in the slide magazine (with drying agent), which is enveloped in a piece of black paper. Allow exposure take place at 4°C for proper time.
f. Restore the slide magazine to room temperature. The prepared developer D19, fixer Kodak F-5 and distilled water cool down to 14°C in the freezer.
g. Develop an image for 4 minutes. Rinse with water for 10 seconds. Fix for 4 minutes. Rinse with water for 20 minutes.
h. Take the sections out of dark room. Clean the sections back side and areas surrounding the tissue.
i. Contrast stain with hematoxylin, Giemsa or methyl green.
j. Perform routine dehydration, clearing and mounting.

Main points
a. Perform radioautography strictly away from light from the beginning to the end.
b. The utensils should be very clean.
c. There are three types of nuclear emulsion: NTB_1, NTB_2 and NTB_3, which are sensitized to α, β and γ rays, respectively.
d. In general, 0.6 mol/L ammonium acetate or water is used to dilute nuclear emulsion (1:1, v/v). Dispense into aliquots and store away from light at 4°C.
e. Strictly control the developing and fixing temperature.

Result observation and quantitative analysis
a. Ideal results: Silver grains in the positive cells are obviously more than background. The silver grains are uniform in size and well distributed. For mRNA sequence of interest, the silver grains are mainly localized in the cytoplasm; for DNA sequence of interest, the silver grains are mainly localized in the nucleus.
b. Quantitative analysis is indicated by the number of silver grains per unit cells or per unit area. Image analyzer is used to count the number of silver grains and estimated area.

13.3.2.4.2 Detection methods for nonradioactive probes

–   Detection with DIG-labeled RNA probes
   a.   Rinse the sections after hybridization to remove nonspecifically bound probes.
   b.   Incubate the sections for 15 minutes with blocking mixture.
   c.   Incubate the sections for 60 minutes with alkaline-phosphatase-conjugated anti-DIG antibody (diluted 1:500 in blocking mixture).
   d.   Rinse the sections three to five times with Tris-buffered saline (TBS).
   e.   Prepare nitroblue tetrazolium (NBT)/5-bromine-4-chlorine-3-3indole phosphate (BCIP) color reagent as recommended by the manufacturer (Mix 45 µL NBT solution and 35 µL BCIP solution in 10 mL detection buffer). This color substrate solution must be prepared freshly.
   f.   In a Coplin jar in the refrigerator, incubate the sections with color reagent until sufficient color develops.
   g.   Stop the color reaction by rinsing the sections several times with tap water.
   h.   Rinse the sections with distilled water.
   i.   Mount the sections directly with any water-soluble mounting medium (GVA Mount).

Note: Incubation can be extended up to 120 hours if the color reagent is replaced every time it precipitates or changes color. Optionally, counterstain the sections with 0.02% bright green or 0.1% nuclear fast red for 1–2 minutes, or use immunocytochemistry.

–   Detection method for biotin probes: same as immunohistochemistry
–   Detection method for alkaline phosphatase: refer to detection with DIG-labeled RNA probes.
–   Solutions preparation
   a.   Blocking solution

10× Blocking stock solution (10%)	5 mL
Fetal bovine serum	5 mL (10%)
Maleic acid buffer	40 mL

   b.   10× Blocking stock solution (10%)

Blocking reagent for nucleic acid hybridization	10 g
Maleic acid buffer	100 mL

The specimen is heated to assist dissolution and sterilized by autoclaving. Dispense into aliquots and store at –20°C.

   c.   Maleic acid buffer (100 mmol/L maleic acid; 150 mmol/L NaCl, pH 7.5)

1 mol/L maleic acid	5 mL
5 mol/L NaCl	1.5 mL
0.1% DEPC-treated H$_2$O	Make up to 50.0 mL

Adjust the pH to 7.5 with NaOH. Sterilize by autoclaving.

d. 1 mol/L maleic acid

    Maleic acid   34.83 g

    0.1% DEPC-treated H_2O Make up to 300 mL

Sterilize by autoclaving.

e. TBS buffer (50 mmol/L Tris-HCl (pH 7.5); 150 mmol/L NaCl)

    1 mol/L Tris-HCl (pH 7.5)   25 mL

    5 mol/L NaCl   15 mL

    0.1% DEPC-treated H_2O 460 mL

f. 1 mol/L Tris-HCl: Dissolve 121.1 g Tris base in 800 mL 0.1% DEPC-treated H_2O. Adjust the pH to the desired value by concentrated HCl.

    pH   HCl

    7.4   70 mL

    7.6   60 mL

    8.0   42 mL

Allow the solution to cool to room temperature before making final adjustments to the pH. Adjust the volume of the solution to 1 liter with 0.1% DEPC-treated H_2O. Dispense into aliquots and sterilize by autoclaving.

g. 5 mol/L NaCl

    NaCl   292.2 g

    0.1% DEPC-treated H_2O Make up to 1000 mL

Dispense into aliquots and sterilize by autoclaving.

h. Detection buffer (100 mmol/L Tris-HCl, 100 mmol/L NaCl, pH 9.5)

    1 mol/L Tris-HCl (pH 9.5)   5 mL

    5 mol/L NaCl   1 mL

    DDW   Make up to 50 mL

13.3.3 Detection of mRNA on paraffin-embedded sections with DIG-labeled RNA probes

13.3.3.1 Tissue preparation (as noted above)

13.3.3.2 Prehybridization treatment

13.3.3.2.1 Procedures

Steps

a. Dewax the sections extensively by treating with xylene (preferably overnight) and a graded series of ethanol solutions.

b. To preserve the mRNA during the following procedure, fix the sections once again with 4% paraformaldehyde (in 100 mmol/L phosphate buffer) for 20 minutes. Rinse the sections three to five times with TBS buffer.

c. Treat the sections for 10 minutes with 200 mmol/L HCl to denature proteins. Rinse the sections three to five times with TBS buffer.

d. To reduce nonspecific background, incubate the sections for 10 minutes on a magnetic stirrer with a freshly mixed solution of 0.5% acetic anhydride in 100 mmol/L Tris (pH 8.0). Rinse the sections three to five times with TBS buffer.

e. Treat the sections for 20 minutes at 37°C with proteinase K (10–500 µg/mL in TBS, which contains 2 mmol/L $CaCl_2$). The concentration of proteinase K depends upon the degree of fixation. Generally, start with 20 µg/mL proteinase K for paraformaldehyde-fixed tissue. The concentration needed for glutaraldehyde-fixed or over-fixed tissue is generally higher than for paraformaldehyde-fixed tissue. Rinse the sections three to five times with TBS buffer.

f. Incubate the sections at 4°C for 5 minutes with TBS (pH 7.5) to stop the proteinase K digestion.

g. Dehydrate the sections in a graded series of ethanol solutions (increasing concentrations of ethanol).

h. Rinse the sections briefly with chloroform.

Note: Do not postfix the sections with paraformaldehyde after protease digestion because this reduces signal intensity. At this point, you may, if necessary, store the sections under dust- and RNase-free conditions for several days.

Perform pretreatment steps under RNase-free conditions. Prepare all solutions and buffers for these steps with DEPC-treated water. Bake all glassware for 4 hours at 180°C. Assume that disposable plastic ware is RNase free.

13.3.3.2.2 Solutions preparation

a. HCl (200 mmol/L)
Concentrated HCl	820 µL
0.1% DEPC-treated H_2O	50 mL

b. Acetic anhydride (0.5%) (make fresh)
Acetic anhydride	250 µL
1 mol/L Tris-HCl (pH 8.0)	5 mL
0.1% DEPC-treated H_2O	44.75 mL

c. Proteinase K working solution (20 µg/mL)
1 mol/L Tris-HCl (pH 8.0)	5 mL
Proteinase K (10 mg/mL)	100 µL
0.5 mol/L EDTA (pH 8.0)	5 mL
0.1% DEPC-treated H_2O	40 mL
0.5 mol/L EDTA (pH 8.0)	
EDTA-Na·2H_2O	186.1 g
0.1% DEPC-treated H_2O	800 mL

Stir vigorously on a magnetic stirrer. Adjust the pH to 8.0 with NaOH (approximately 20 g NaOH). Adjust the volume to 1000 mL. Dispense into aliquots and sterilize by autoclaving.

13.3.3.3 Hybridization and rinse

13.3.3.3.1 Hybridization
Steps
a. Place the sections in a humid chamber at 55°C for 30 minutes.
b. Prepare a hybridization buffer.
c. Dilute the labeled antisense RNA probe to an appropriate degree in hybridization buffer (the working concentration of antisense RNA probe is 0.2–10 ng/µL). The diluted probe solution may contain as much as one part labeled probe to four parts hybridization buffer.

Note: The amount of probe needed depends on the degree of probe labeling. Generally, use the lowest probe concentration that gives optimal response in that dot blot procedure.

Example: In a serial dilution of the probe on a dot blot, if 1:160 dilution gives a significantly stronger response than 1:320 dilution, perform the initial hybridization experiments with both 1:200 and 1:300 dilution of the probe.

d. Pipette the diluted antisense RNA probe solution onto each section at a volume of 10 µL/cm². Cover with a coverslip. Use the rubber cement to seal the coverslip.
e. As a control, serve sections treated in the same way whereby a labeled sense RNA probe is used in steps c and d.
f. Place the sections on a hot plate at 95°C for 4 minutes.

Note: This step increases the signal from RNA/RNA hybrids.

g. Incubate the sections in a humid chamber for 4–6 hours at 55°C–75°C.

Note: Hybridization specificity can be increased by increasing the temperature of hybridization, if necessary, as high as 75°C. Increasing the temperature can help to differentiate mRNAs of highly homologous proteins.

Note: Perform hybridization steps under RNase-free conditions. Prepare all solutions and buffers for these steps with DEPC-treated water. Bake all glassware for 4 hours at 180°C. Assume that disposable plastic ware is RNase free.

13.3.3.3.2 Posthybridization rinse
Note: Stringency rinse, often described as a method to remove nonspecifically bound probes, is helpful in reducing diffuse background staining but does not significantly improve the specificity of the hybridization.

a. Remove the rubber cement. Incubate the sections in 2× SSC overnight. The coverslips will float off during this incubation.
b. Rinse the sections as follows:
   i. 20 minutes, thrice, at 55°C with 50% deionized formamide in 1× SSC.
   ii. 15 minutes, twice, at room temperature with 1× SSC.
c. Rinse the sections three to five times with TBS.

13.3.3.3.3 Solutions preparation
a. Hybridization buffer

20× SSC	50 μL (2× SSC)
50% dextran sulfate	100 μL (10%)
10 mg/mL sheared salmon sperm DNA	5 μL (0.01%)
1% sodium dodecyl sulfate (SDS)	10 μL (0.02%)
100% deionized formamide	250 μL (50%)
0.1% DEPC-treated H_2O	85 μL

Note: The concentration of dextran sulfate is critical. Without appropriate amounts of dextran sulfate, the method loses sensitivity. However, excessive concentration of dextran sulfate causes higher viscosity of the hybridization buffer. To prevent uneven distribution of the probe and uneven signals, always vortex the hybridization buffer extensively.

b. 10% SDS

Electrophoresis-grade SDS	100 g
0.1% DEPC-treated H_2O	Make up to 1000 mL

The specimen is heated to 68°C to assist dissolution. If necessary, adjust the pH to 7.2 by a few drops of concentrated HCl. Store at room temperature. Sterilization is not necessary. Do not autoclave.

13.3.3.4 Detection of mRNA
As in procedure before, the positive staining appears as black-blue in color, as shown in Fig. 13.3.

13.4 Factors affecting *in situ* hybridization

13.4.1 Concentration of probe

Concentrations of probe directly affect the speed of hybridization. Lower concentrations of probe lead to slower hybridization reaction; conversely, higher concentrations of probe lead to quicker hybridization reaction. The working concentration (radioisotope probe, 0.5 ng/μL; nonradioactive probe, 2 ng/μL) is suggested.

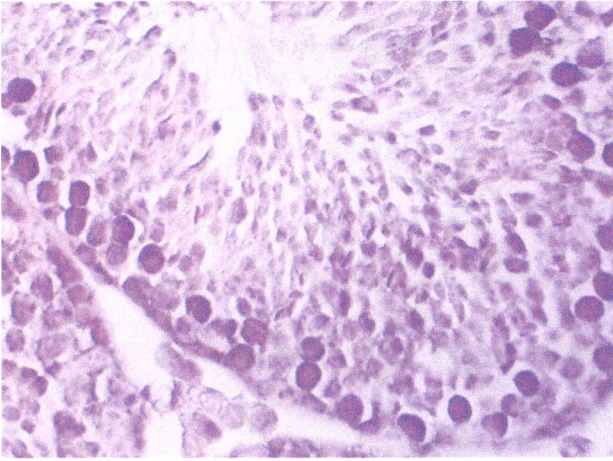

Fig. 13.3: *In situ* hybridization with DIG-labeled antisense RNA probes shows that specific Tex11 mRNA locate in the cytoplasm of spermatogonia and primary spermatocytes in adult mouse testis.

13.4.2 Hybridization temperature

Hybridization temperature should be 20°C–30°C lower than the unwinding or melting temperature (T_m) of hybrids. T_m is the temperature at which half hybrids uncoil to form single strands. For example, if T_m of hybrids is 70°C, then the hybridization temperature should range from 40°C to 50°C. The T_m of RNA-RNA hybrids is the highest, followed by the DNA-RNA hybrids. The T_m of DNA-DNA hybrids is the lowest. The difference between them is approximately 5°C.

13.4.3 Hybridization time

Hybridization time may shorten with high concentrations of probe, but in general, the hybridization reactions complete within 4–6 hours. In practical operation, you may incubate the sections in hybridization buffer overnight.

13.4.4 Hybridization buffer

Hybridization buffer contains not only labeled probe but also salt, formamide, dextran sulfate, BSA and carrier DNA or RNA. Higher concentrations of Na$^+$ can increase hybridization efficiency and decrease static electricity binding of probe to tissues. In general, 0.3–0.9 mol/L NaCl is adopted. Formamide can lower T_m. Formamide (1%) in hybridization buffer lowers RNA-RNA, RNA-DNA and DNA-DNA hybridization temperature at 0.35°C, 0.5°C and 0.65°C, respectively. Thus, 50% formamide is used to avoid the destruction of tissue morphology and specimen falling off caused by too high hybridization temperature. Dextran sulfate combines with H_2O, hence decreasing the volume of hybridization buffer and increasing the effective probe concentration to boost hybridization efficiency. BSA and carrier DNA or RNA (e.g., salmon sperm DNA and *Bacillus coli* tRNA) in hybridization buffer are used to block the nonspecific binding of probe to tissues, thus reducing background staining. The pH variation does not affect hybrid formation if the pH level of the hybridization buffer is 5.0–9.0. Usually, the pH level of the hybridization buffer is 6.5–7.5; hybridization buffer contains 20–50 mmol/L phosphate. If the ^{35}S-labeled probe is used, DTT should be added to the hybridization buffer, and its final concentration is 100 mol/L.

13.5 Control test

To prove the accuracy of *in situ* hybridization experimental operation and the specificity of the experimental results, it is necessary to set up a series of positive and negative controls such as tissue control, probe control, hybridization reaction control and detection system control.

13.5.1 Tissue control

13.5.1.1 Northern or Southern blot
Positive results of *in situ* hybridization demonstrate the presence of target DNA or RNA *in situ* within the tissue cells. DNA or total RNA is extracted from the same tissues, then Southern or Northern blot is performed.

13.5.1.2 Immunohistochemistry
If the antibody against the target gene product is available, immunohistochemistry stain can be performed with the same specimen. The positive results of immunohistochemistry support the positive results of *in situ* hybridization.

13.5.1.3 Perform *in situ* hybridization with actin-RNA probe or human placenta DNA probe
To prove well preservation of RNA or DNA in the specimen during the preparation process, perform *in situ* hybridization on the tissue sections with actin-RNA probe or human placenta DNA probe. Because human tissue cells generally contain actin mRNA and the complementary DNA sequence with human placenta DNA probe, positive results should be obtained.

13.5.2 Probe control

13.5.2.1 Known positive tissue and negative tissue control
Perform *in situ* hybridization with probe on the known positive tissue, which contains the target nucleotide sequence, and known negative tissue, which does not contain the target nucleotide sequence. Then positive and negative results should be obtained respectively.

13.5.2.2 Use labeled carrier as probe
When labeled probe contains carrier sequence, the positive results indicate not only possibly specific probe-target-nucleic-acid hybridization signal but also possibly nonspecific binding of carrier sequence to the specimen nucleic acid. You may perform *in situ* hybridization with labeled-carrier probe, and the negative results rule out nonspecific binding of carrier sequence to the specimen nucleic acid.

13.5.2.3 Perform *in situ* hybridization with labeled sense RNA probe
This is a good negative control to prove the specificity of probe.

13.5.3 Hybridization reaction control

13.5.3.1 Hybridization with absent probe
The *in situ* hybridization results should be negative with absent labeled probe. This is a convenient and significant negative control. The positive reactions with absent labeled probe indicate false-positive signal.

13.5.3.2 Prehybridization DNase or RNase treatment
Use DNase or RNase to treat the specimen according to corresponding target DNA or RNA nucleic acid, and then perform *in situ* hybridization. However, this control test cannot prove specific probe and target DNA or RNA hybridization. Be sure to remove the enzyme on the specimen thoroughly after prehybridization DNase or RNase treatment.

13.5.3.3 Competition test between labeled probe and nonlabeled probe
Mix the labeled probe and nonlabeled probe with a certain gradient ratio, then perform *in situ* hybridization. The hybridization signal should weaken with higher concentrations of nonlabeled probe in case of specific hybridization reaction.

13.5.4 Detection system control

13.5.4.1 Radioautography detection control

Radioactive isotope autograph controls include positive and negative controls of blank film. The former exposes the blank film immersed in foam rubber to the light allowing developing an image, and positive results indicate the normal foam rubber and developing work. The latter carries out radioautography with the blank film after immersing in foam rubber and *in situ* hybridization specimen; the results should be negative. If numerous silver grains appear in the comparison film, this illustrates the radiocontamination of foam rubber, and you should use new foam rubber.

13.5.4.2 Nonradioactivity *in situ* hybridization detection system control

Immunohistochemistry stains serve as the controls, which include a series of positive and negative controls of immunohistochemistry.

Knowledge links: The fluorescence image

The fluorescence *in situ* hybridization (FISH) technology for RNA detection in alive cells has been reported. This detection depends on nature fluorescence group in the body or artificial fluorescence, which was labeled on probes. Compared with green fluorescence protein (GFP), it is easier to detect different target molecules, but it is also easy to be affected by cell circumstance. The combination of FISH technology and GFP can be used to examine the target nucleic acid and protein.

The appearance of multiphoton microscopy technology increases the application of fluorescence images because its electronic module could emit laser two to three times to stimulate the fluorescence as required. If far infrared ray is introduced, the detection can be operated for deep-inside structures in tissue and even in the whole body with very low toxicity, and this is new method to produce fluorescence image for whole and alive animal. If body probe is applied in the future, it is possible to identify the specific nucleic acid or its invasion in the body.

14 *In situ* polymerase chain reaction histochemical technology

In situ polymerase chain reaction (ISPCR) technology was first reported in 1990. It is easy for highly sensitive ISPCR to detect virus DNA with low copies and almost any other nucleic acid sequences, even the gene sequence of only one copy.

14.1 Basic theory

ISPCR somehow conjugates PCR and *in situ* hybridization together; thus, it is used to detect very low level of DNA or mRNA location directly in tissue sections. ISPCR takes advantage of the powerful amplification capacity of PCR to amplify the target gene, and the PCR products remain in cells because of their large molecules or interlacing structures. Then the PCR products are detected by *in situ* hybridization with labeled probe. This technique possesses the advantages of high sensitivity, specificity and precise cell localization.

14.2 Basic types

14.2.1 Direct ISPCR

In the direct method, the labeled nucleotides or primers are incorporated in the PCR product and subsequently detected.

14.2.2 Indirect ISPCR

PCR amplification of DNA is first performed, and then *in situ* hybridization is used with labeled probe. This method is also called PCR *in situ* hybridization.

14.2.3 *In situ* reverse transcription PCR (RT-PCR)

In situ RT-PCR applies liquid-phase RT-PCR to tissue sections. The tissue specimens are treated with DNase before *in situ* RT-PCR reaction to ensure that the amplified templates come from cDNA. Other steps are similar to that in liquid-phase RT-PCR.

DOI 10.1515/9783110531398-014

14.3 Procedures

The basic procedure of ISPCR technique includes tissue preparation, ISPCR and *in situ* detection. These steps are described in the following sections (e.g., paraffin section).

14.3.1 Tissue preparation

The ISPCR technique can be applied to cell suspension, cell smear, frozen section and paraffin section. Performing ISPCR with the suspended intact cells obtains better results, but the results are worse for paraffin sections.

a. Fixation. Use 10% formalin or 4% paraformaldehyde as fixatives.
b. Section thickness. The thicker the section, the better the results ISPCR can achieve. However, thicker sections affect tissue morphology.
c. Section treatment. The purpose of slide treatment is to prevent tissues from shedding off.
d. Digestive effect of protease. Protease treatments increase the permeability of cells and the exposure of target sequence for amplification. The often-used proteases include protease K, trypsin and pepsin. The specimen is heated to inactivate proteases after proteases digestion or to remove the proteases by sufficient rinse.

14.3.2 ISPCR

14.3.2.1 Primer
Shorter primers (15–30 bp) are generally used, and the amplified product length is 100–1000 bp.

14.3.2.2 Reaction system
The reaction system of ISPCR is basically identical with the conventional liquid-phase PCR. Because ISPCR is performed on the fixed sections, the concentrations of primers, Taq DNA polymerase and Mg^{2+} should be higher than that in the liquid-phase PCR reaction system.

14.3.2.3 Heat cycle
Perform heat cycle of ISPCR on the specialized thermal cycler. To prevent loss of reaction system, you may need to seal the coverslips.

14.3.2.4 Rinse
Rinse the specimen after PCR amplification to remove the amplified products that disperse out of the cells. Insufficient rinse will lead to strong background or false-positive result.

14.3.3 *In situ* detection

The detective method of the amplified products is determined by the design proposal of ISPCR. In case of direct method, perform *in situ* detection according to the property of labeled molecules. As to indirect method, the detection is carried out with *in situ* hybridization. If fluorescence, biotin, alkaline phosphatase and horseradish peroxidase are used as markers, the detections are identical with those in immunohistochemistry.

14.3.4 Procedure of direct ISPCR

a. Tissue sections (4–10 μm thick). Bake the sections at 60°C for 90 minutes. Cool down to 20°C for 24 hours and keep at room temperature.
b. Dewax paraffin sections by treating with xylene and a graded series of ethanol solutions. Rinse the sections with 0.1 mol/L Tris-HCl (Tris buffer, pH 7.4) for 5 minutes.
c. Protease digestion
d. Dilute protease K with 0.1 mol/L Tris-HCl (pH 8.0) and 10 mmol/L ethylenediaminetetraacetic acid. (The concentration of protease K is 10 μg/mL for tissue sections and 20 μg/mL for cell slides). The digestive time depends on protease concentrations and tissues, usually at 37°C for 20 minutes. Place the sections at 96°C for 2 minutes to inactivate protease. Rinse the sections with 0.1 mol/L TB (pH 7.4).
e. Add 100 μL 1× Taq buffer onto the sections. Cover the specimen with coverslip, and undergo denaturation at 95°C for 5 minutes.
f. Prepare amplification mixed liquor (30 μL per slide): 1× Taq buffer, 200 μmol/L dNTP (with biotin or DIG-labeled dATP), 100 pmol/L primer and 5 U Taq enzyme. Make up to 30 μL of total volume with deionized water.
g. Pipette 30 μL of amplification mixed liquor onto each section. Cover with a coverslip. Use nail polish to seal the coverslip, and then perform PCR. PCR reaction parameters are designed according to the primers and amplified fragments. In general, 20 cycles are needed.
h. Detection of PCR products
    i.   Remove the coverslips. Rinse the sections with TB for 5 minutes, twice.
    ii. Rinse the sections with iced ethanol and 80% ethanol, respectively, for 5 minutes, twice.

iii. Incubate the sections with streptavidin-AKP 1:100 or anti-Dig-AKP 1:500 at 37°C for 30–60 minutes.
iv. Rinse the sections with TB for 5 minutes, twice. Perform coloration.
v. Stop the reaction and mount the sections.

14.3.5 Procedure of indirect ISPCR

a. ISPCR amplification is identical with the direct method, but the dNTPs are not labeled.
b. Perform *in situ* hybridization after PCR amplification.

Prepare 2× saline sodium citrate (SSC) hybridization solutions with DIG-labeled probe (5 ng per section, 20 μL per section). Before hybridization, denature probe at 95°C for 10 minutes. Pipette the probe solution onto each section. Incubate the sections in a humid chamber overnight at 42°C.

c. Remove the coverslips. Rinse the sections with 4× SSC for 2 minutes at room temperature.
d. Rinse the sections with 2× SSC, 1× SSC, 0.5× SSC and 0.2× SSC, respectively, at 42°C for 10 minutes, twice. Rinse the sections with TB for 5 minutes.
e. Incubate the sections with anti-Dig-AKP 1:500 at 37°C for 2 hours.
f. Perform NBT/BCIP coloration in a dark place for 30 minutes to 2 hours.
g. Rinse the sections to stop coloration. Counterstain with nuclear fast red or methyl green for 5 minutes and mount the sections.

14.3.6 Control test

a. Taq DNA polymerase is removed to determine the presence or absence of nonspecific adsorption of probe or antibody.
b. Primers are removed to rule out the phenomenon of DNA artificial repair or endogenous primer.
c. RNase or DNase pretreatment.
d. Substitute independent probe or antibody in *in situ* hybridization.
e. In case of negative ISPCR amplification, tissue or cell DNA is extracted to perform liquid-phase PCR.

14.4 Application of ISPCR technology

The predominance of ISPCR technology is that specific gene sequence of low copy number can be detected *in situ* in cells.

14.4.1 Detection of exogenous gene

14.4.1.1 Detection of viral gene
Good test methods are not available to detect cells infected by virus until now. ISPCR technology can be used to solve this problem.

14.4.1.2 Detection of bacterial gene
If the tuberculosis is not typical, it is very difficult to find bacillus tuberculosis stained by a specific stain under the light microscope. ISPCR technology can be used to assist the final diagnosis.

14.4.1.3 Detection of imported gene
ISPCR technology can be used to confirm the imported genes in transgenic animal researches and in patients that are treated by gene therapy.

14.4.2 Detection of endogenous gene

14.4.2.1 Detection of abnormal gene
ISPCR technology can be used in tumor researches and diagnosis. This technology is used to detect gene mutation of proto-oncogene, anti-oncogene and gene rearrangement, such as rearrangement of malignant lymphoma immunoglobulin heavy chain gene.

14.4.2.2 Detection of intrinsic gene
It is helpless for *in situ* hybridization technology to detect low expressed intrinsic genes with single or only several copies. Although liquid-phase PCR can detect these genes, cell types that express the genes cannot be determined. ISPCR technology can solve this problem.

Knowledge links: PCR technology and its development

In 1985, American scientist Kary Mullis invented PCR technology and published the research paper on *Science*. Since then, more and more researchers accepted PCR technology, and Kary Mullis won the Chemistry Nobel Prize in 1993.

The study for nucleic acid has been conducted for more than 100 years. In the 1960s and 1970s, scientists focused on the gene *in vitro* separation technology, but they could not make a breakthrough because of the limited amount of DNA. In 1971, Khorana proposed the hypothesis of gene amplification *in vitro*, but because the technology of gene sequence analysis was immature, the heat-resistant DNA polymerase could not be found and the synthesis of oligonucleotide primers was semiautomatic. This hypothesis had no practical significances. In 1988, Keohanog improved the PCR technology by enhancing enzyme efficiency.

Finally, Saiki extracted heat-resistant DNA polymerase from aquatic thermophilic bacillus in a hot spring in Yellowstone National Park in the United States. This greatly enhanced PCR technology and facilitated the wide application of PCR. PCR became the popular and basic technology for mordent molecular biology. After that, PCR technology was improved and reformed accordingly; for example, PCR can now be used from qualitative assay to quantitative assay; not only few kb genes but also few decades of kb genes can be amplified; RT-PCR is established if reverse transcriptase is introduced in PCR technology; and immuno-PCR technology is created if immunohistochemistry combines with PCR technology.

15 Electron microscopic histochemistry technology

Electron microscopic cytochemical technology, also known as ultrastructural cytochemistry technology, is a technology that combines histochemistry with electron microscopy technology and has been applied in cytological study. It has taken the development of histochemical study from microscopic level to ultramicroscopic level. The theory of the technology is that specific chemical reactions produce specific reaction products and then, in turn, form high electron density insoluble sediments. Finally, by analyzing the electron densities of the sediments under an electron microscope, the intracellular chemical compositions will be ultrastructurally located, and the activities of the compositions are then evaluated. According to different detection targets and methods, the electron microscope cytochemistry technology can be divided into electron microscopic enzyme cytochemistry technology and electron microscopic immunocytochemistry technology.

15.1 Electron microscopic enzyme histochemistry technology

Electron microscopic enzyme histochemistry technology aims to detect intracellular enzymes (proteins) by using an electron microscope. It is mainly used to research the marker enzymes of organelles and cell membrane system. Currently, more than 80 categories of marker enzymes can be displayed through electron microscopic cytochemistry methods. Because of the accurate and qualitative positioning of various kinds of enzymes and the increasing detection rates, the electron microscopic enzyme cytochemistry technology has been widely used in scientific research to reveal the relationships between the structure and the function of all kinds of cells.

15.1.1 Procedures of electron microscopic enzyme histochemistry

The principles of electron microscopic enzyme histochemistry are similar as that of light microscopy and are summarized as below.

15.1.1.1 Sampling
The process of sampling demands accurate positioning and directionality of the specimen. The size of the samples should be in 0.5–1.0 mm^3. The samples must be treated by 0–4°C fixative solutions as quick as possible.

15.1.1.2 Fixation
In electron microscopic enzyme histochemistry, the purpose of the fixation is not only to preserve the complete ultrastructures but also to save the enzyme activities and

DOI 10.1515/9783110531398-015

accurate positions. Thus, the types of fixatives, concentrations, pH values, time span and temperature are of great importance. Newly removed unfixed fresh tissues within 5 minutes, which still maintain high enzyme activity, are cut into 10–40 μm thick sections for quick incubations usually bring out satisfactory results.

The commonly used fixatives in electron microscopic enzyme histochemistry are glutaraldehyde and paraformaldehyde. Besides, both osmium tetroxide and potassium permanganate are also used as fixing agents. However, they are strong oxidizers and may destroy the activities of enzymes and cannot be used singly.

Preparation for paraformaldehyde fixative solutions
Paraformaldehyde	4 g
0.1 mol/L dimethylarsinate buffer (pH 7.2–7.4)	100 mL
4.0% formaldehyde was often used as well.	

Preparation for glutaraldehyde fixative solutions
25% glutaraldehyde	16 mL
0.1 mol/L dimethylarsinate buffer, pH 7.6	84 mL
Sucrose	8 g

Preparation for paraformaldehyde and glutaraldehyde mixed fixative solutions
4% paraformaldehyde	92 mL
0.1 mol/L dimethylarsinate buffer, pH 7.2	100 mL
25% glutaraldehyde	8 mL

The dual fixation method has been commonly adopted in the electron microscope cytochemistry, called prefixation and postfixation. Prefixation is used to maintain the ultrastructure and to prevent the enzyme activity from being destroyed, in which the samples are fixed for 1–2 hours at 4°C, and postfixation is used to preserve the ultrastructure and to enhance the electron density well. Usually, 1% osmium tetroxide (OsO_4) is used to fix the samples for 30–60 minutes on ice or at 4°C.

15.1.1.3 Rinse
After prefixation, the samples must be fully rinsed by fixation medium buffer before entering the incubation solutions.

15.1.1.4 Presection
Before incubation in electron microscopic enzyme cytochemistry, samples should be presectioned, and after that, incubation reaction is conducted. Presectioning can facilitate the penetration of incubation medium into cells and tissues. In the process of presectioning, the fully rinsed samples are sectioned into 40–60 μm thick slices in freezing microtome, cryostat microtome or vibratome, of which the vibratome is recommended. After incubation reaction, the presectioned thin slices will be ultrathin sectioned.

15.1.1.5 Incubation reaction

The process of incubating the presectioned tissue slices into selected incubation solution that enables specific chemical reactions to occur is known as incubation reaction, or positioning reaction. It is the key program of the electron microscope cytochemistry.

The preparation for incubation buffer has the following strict requirements:

a. The chemical reagents used must be first class (GR), and laboratory wares must be absolutely clean.
b. The reagents should be added according to sequences of the incubation formula; generally, the sequence is not allowed to change so as to avoid the changes of the solution concentrations and the effects of incubation reaction.
c. Continually vibrate or stir when joining in the new regents in sequence to make it uniform and fully mixed.
d. Determine the pH value after preparation.
e. The concentrations of the substrates must be accurate.

The incubation methods are as follows:

a. Slide incubation method: The tissue slice is first adhered to a glass slide or coverslip and then immersed in incubation buffer. This method is particularly suitable for studies of unfixed samples, free suspension cells and cultured cells.
b. Floating incubation method: The tissue slice is directly immersed in the incubation buffer; 10% (v/v) sucrose or dimethyl sulfoxide (DMSO) can improve the permeability of incubation buffer. This method is particularly suitable for studies of unfixed samples, free suspension cells and cultured cells.

The factors that affect electron microscopic enzyme cytochemistry are basically the same as that in the enzyme histochemistry. However, factors such as reaction temperatures, incubation time span, sample types and thickness and pH values of incubation solutions should also be noticed.

15.1.1.6 Rinse

After incubation reaction, the sample is rinsed in the same buffer in which the incubation buffer is prepared for 20–60 minutes. Replacing the buffer two to three times will help rinse the reagents, leaving in the tissue samples, and it will benefit the subsequent postfixation.

15.1.1.7 Postfixation

After incubation, the sample is fixed with 1% osmium tetroxide (OsO_4) at 4°C or 30–60 minutes on ice. The postfixation is able to preserve cell ultrastructure well and enhance the electron density, which would increase the image contrast under EM.

15.1.1.8 Rinse
Rinse the sample in 0.1 mol/L sodium dimethylarsinate buffer two to three times, 10 minutes each time, to remove redundant osmium tetroxide in sample.

15.1.1.9 Dehydration
The sample is dehydrated with ethanol or acetone. This will help in the soaking of the embedding medium. Dehydration must adopt the "grade series dehydration method" to avoid cell shrinkage. Commonly, the dehydration solution should be 30%, 50%, 70%, 80%, 90% and 100% ethanol or acetone. Please note the sample in each concentration for 10–15 minutes.

15.1.1.10 Embedding
At room temperature, the sample is embedded by immersion into the following soaking solutions:
a. Epon812: acetone = 1:3   2 hours
b. Epon812: acetone = 1:1   2 hours
c. Epon812: acetone = 3:1   2–4 hours
d. Soaking in pure Epon812 embedding medium 1–2 days

During soaking, tilted rotational vibration or vacuuming can be adopted to improve the efficiency of penetration. After soaking, the sample is embedded in newly prepared Epon812 embedding medium with ventilation. The formula of embedding medium is as follows:
a. Epon812   13 mL
b. Dodecylsuccinic anhydride (DDSA)   8 mL
c. Methyl nadic anhydride (MNA)   7 mL
d. 2,4,6-Tri (dimethylaminomethyl) phenol) (DMP-30)   10–12 drops

According to the sample situation and research requirement, the embedding board or medical capsules can be used for embedding orientation. The embedded sample is immersed into incubator for polymerization. The embedded media are polymerized for 12, 12, 24 hours at 37°C, 45°C and 60°C, respectively, or polymerized at 60°C for 48 hours.

15.1.1.11 Ultrathin section
After embedding, strip the capsule, locate and mend the tissue sample under anatomic microscope. After mending, the top surface of the sample is usually 0.5 mm × 0.5 mm. The sample is then ultrathin sectioned, and the tissue should be controlled evenly in thickness (in 50–70 nm generally) and should successively form a tissue slice belt. Then the slices are floated on the surface of the water tank. After that, filter papers dipped with chloroform or xylene are used for slice flatting. During slice flatting, slowly move the paper on the water surface of the tank and approach the slice belt and hold the paper for 3–4 seconds to help slice flatting. After slice flatting, micro eyelash is used to

cut the floating slice belt into several sections, four to five pieces in each section. After that, clamp the grid (with support membrane) using tweezers, and then salvage the slices from the bottom side. After the slices are loaded on the grid, absorb the excess water around the grids carefully by filter paper, and the slices are naturally dried for subsequent electronic staining (poststaining).

15.1.1.12 Electronic staining

Electronic staining is also called poststaining. Before poststaining, the sections must be observed first. If the electron densities of ultrathin sections have an obvious contrast, then it needs no further electronic staining. If the contrast is not obvious, conduct electronic staining. In electronic staining, the sample is first stained with 50% ethanol acetate-saturated solution of uranium (uranium acetate 2 g dissolved in 50% ethanol 100 mL) for 20–30 minutes. Then the sample on the grid is rinsed by double distilled water (DDW) three times. After the rinse, the sample is blotted up by filter paper, naturally dried and observed under an electron microscope.

15.1.1.13 Electron microscope observation

Under an electron microscope, the ultrastructures of cells, the electron densities of productions during chemical reactions and its distributions should be carefully observed. A well-stained sample should be uniform in fine particles. The reaction product positions of marker enzymes in each organelle ought to be the locations of the corresponding organelles.

15.1.2 Examples of electron microscopic enzyme histochemistry

15.1.2.1 Alkaline phosphatase

Alkaline phosphatase (AKP) is a kind of enzyme that catalyzes the hydrolysis reaction of various phosphates of alcohol and phenol in alkaline environment. The lead citrate method (one of metallic-salt method) is always used in detecting AKP in electron microscopic enzyme histochemistry (EMEC). This method has the following advantages: First, it can be used in both light microscope (LM) and electron microscope (EM) level. Second, lead citrate is stable under alkaline environment; thus, the incubation solution is also transparent and stable. Third, the reaction product is fine and suitable for observation under an electron microscope. Fourth, as long as the substrates are changed in the reaction, the method can be used in different AKP activity detection. The main procedures are as follows:

Pretreatment

a. Fixation: 2% glutaraldehyde (0.1 mol/L dimethylarsinate buffer, 8% sucrose, pH 7.2–7.4) for 60 minutes or 4% paraformaldehyde for several to 12 hours.
b. Sectioning: The tissue is cut into 40–50 μm thick sections by freezing microtome or vibratome.

Preparation of the incubation solution

0.2 mol/L Tris-HCl buffer (pH 8.5)	1.4 mL
0.1 mol/L sodium β-glycerophosphate	2.0 mL
0.015 mol/L MgSO$_4$	2.6 mL
Sucrose	0.8 g
0.5% lead citrate	4.0 mL

Staining procedures
a. Fresh tissue is prefixed and frozen presectioned, 40 μm thick.
b. Immerse the sections into incubation fluid; incubate for 15 to 30 minutes at 37°C.
c. Postfix the sections in cold 1% osmic acid (OsO$_4$) buffer for 30 minutes.
d. Dehydrate the sections in 95% alcohol.
e. Embed the sections in epoxy resin.
f. The sections are ultrathin sectioned.
g. The sections are electronically stained.
h. Observe under electronic microscope.

Results and evaluation: Lead phosphate presents high electron density particles, which mainly distribute on cell membranes.

Control test: Remove substrates from the incubation solution or add inhibitors (e.g., β-phenylalanine or tetrazolium or bromide tetrazolium) in the reaction system; the enzyme activity would be negative. Add inhibitors (e.g., β-phenylalanine or tetrazolium or bromide tetrazolium); the enzyme activity would be negative. However, inhibitors with strong chelation reactions are not suggested to be chosen because the chelation reaction can not only chelate the Zn^{2+} in AKP molecular structural components but also activate the Mg^{2+}. It can also chelate the capture agent (Pb^{2+} for instance) of free phosphate group, which would cause false negative results.

15.1.2.2 Succinate dehydrogenase

15.1.2.2.1 Theory: tetrazolium salts method
Cupric ferrocyanide method (hexacyanoferrate reduction method): Hexacyanoferrate can receive the free hydrogen from dehydrogenase reaction and can be reduced to ferrocyanide, which deposits in the active sites of the enzymes and would be easily detected because of its high electron densities.

15.1.2.2.2 Tetrazolium salts method
The tetrazolium salts used in electron microscopy include Os-TNST, Tc-NBT, Ds-NBT, tetrazolium orange, 2-(2'-benzothiazolyl)-5-styryl-3-(4'-phthalhydrazidyl) tetrazolium chloride (BSPT), etc., in which the BSPT is mostly used.

Preparation for the incubation solution

0.1 mol/L Tris-HCl buffer (pH 7.2)	5 mL
BSPT (dissolved in 0.25 mL DMSO)	25 mg

Sodium succinate	81 mg
Sucrose	0.5 g
Distilled water	5 mL

Staining procedures
a. Fixation, rinse and sectioning: same as the steps of tetrazolium salt method in LM.
b. Incubate the sections at 37°C, 20–90 minutes.
c. Rinse the sections with Tris-HCl buffer, 10 minutes.
d. Postfix the sections with 2% osmium tetroxide solution for 2–3 hours.
e. Cut ultrathin sections according to the electron microscope requirements, and observe them under the electron microscope.

Results: The reaction products distribute in the outer side of the mitochondrial inner membrane (crista side). Tetrazolium salt method can not only be applied to succinate dehydrogenase when changing the substrates but also be used in detecting other oxidoreductase activities, such as G-6-PDH.

Control test: Remove the substrate or add 10 mmol/L malonate (competitive inhibitors) into incubation solution.

15.1.2.2.3 Cupric ferrocyanide method
Preparation of the incubation solution

Sodium succinate	100 mg
0.1 mol/L phosphate buffer (PB) (pH 7.0)	13 mL
0.1 mol/L sodium citrate	0.6 mL
30 mmol/L copper sulfate	2 mL
5 mmol/L potassium ferrocyanide	2 mL
Sucrose	2–3 g
Distilled water (pH 7.0)	20 mL

Staining procedures
a. Fixation, rinse and section: The same as tetrazolium salt method in LM.
b. Incubation: 30–45 minutes at room temperature.
c. Treatments after incubation: Rinse with precooled buffer for approximately 30 minutes, postfix with osmium tetroxide for 10 minutes, dehydrate and embed with epoxy resin, perform ultrathin sectioning and lead staining and observe under an electron microscope.

15.1.2.3 Results and evaluations
Cupric ferrocyanide with high electron density deposits in the external cavity and crista of the mitochondria (lacuna between inner membrane and outer membrane) and disperses. Succinate Dehydrogenase (SDH) is a kind of flavoprotein that exists in the inner mitochondrial membrane. Sometimes reaction products appear in the synapse membrane or within the Golgi complex membrane capsule. To enhance the enzyme reaction

and to shorten the reaction time, 3%–5% DMSO is always added in incubation solution. However, the DMSO is alkaline and may cause the diffusions phenomenon of the reaction products; therefore, after incubation, completely rinse it to remove the reaction solutions.

15.1.3 Electron microscopic enzyme histochemistry of organelles

Each organelle has several high specific enzymes as the marker enzymes of the organelles. Electron microscopic cytochemistry of organelle enzymes is able to observe the cell chemistry reaction *in situ* and functionally locate the organelles without destroying the cell ultrastructure. Therefore, it is significant to the research of cell ultrastructure and its relationship to cell functions.

15.1.3.1 Lysosome

Lysosome enzymes are mostly hydrolyses, and optimum pH values are in acid range. Among them, the acid phosphatase (ACP) enzyme is widespread in lysosomes, and the enzyme detection method is stable and reliable. Thus, it has been regarded as the marker enzyme of lysosome.

Composition of incubation fluid

0.05 mol/L acetate buffer (pH 5.0)	50 mL
Lead nitrate	4 g
3% β-sucrose glycerin sodium sulfate	5 mL

Staining procedures
a. Fixation: The sample undergoes rapid perfusion fixation by 300–500 mmol/L hypertonic solution for 5–10 minutes, or immersed with 2% glutaraldehyde and 2% paraformaldehyde for 1–2 hours. The fixative is prepared with 0.1 mol/L dimethylarsinic acid salt buffer.
b. Rinse: Rinse the sample with buffer for 1 hour or overnight. Adding a few DMSO in dimethylarsinic acid sodium buffer would enable the incubation medium substrates and capture agents quickly entering into the cells and, thus, strengthen the activity of ACP enzyme. The effects of DMSO are even more remarkable in cultured cell lysosomes in which ACP enzyme activities are not easily detected.
c. Section: Tissue is sectioned into 20–40 μm thick slices by nonfreezing microtome or freezing microtome.
d. Incubation: Soak the sections in incubation solution for 10–60 minutes.
e. Rinse: Rinse the sections with 0.1 mol/L dimethylarsinic acid sodium buffer for 1–2 minutes.
f. Postfixation: Immerse the incubated sections into precooled 1% osmium tetroxide solution for 30 minutes.
g. Final treatment: Acetone dehydration, resin embedding, ultrathin sectioning, electronic staining and observation under an electron microscope.

Results: Lysosome and autophagy-lysosome packing membranes are all positive, whereas phagosome and autophagy bodies are negative.

Note: The ACP enzyme itself is much more tolerant to the fixative, but the unit membrane of lysosome is fragile. Thus, the high permeability (300 mmol/L) fixatives are usually adopted for rapid (5–10 minutes) perfusion fixation.

15.1.3.2 Mitochondrion

Mitochondrion is the power-supplying organ. It contains a lot of redox enzymes, including its marker enzymes, the succinodehydrogenase, the cytochrome oxidase and the succinate dehydrogenase. In the electron microscope cytochemistry, metal salts method has been much used in displaying the succinate dehydrogenase.

Composition of incubation medium

Sodium succinate	100 mg
0.1 mol/L PB (pH 7.0)	13.0 mL
0.1 mol/L sodium citrate	0.6 mL
30 mmol/L copper sulfate	2.0 mL
Distilled water	2.4 mL

After they are completely dissolved, add the following:

5 mmol/L potassium ferricyanide	2.0 mL/L
Sucrose	3.0 g

Staining procedures
a. Fixation: The sample is fixed in 0.25%–0.3% glutaraldehyde or 0.5% formaldehyde for 10–20 minutes.
b. Rinse: Rinse the sections adequately by buffer.
c. Section: The sample is presectioned into 30–40 μm thick slices.
d. Incubation: The incubation is taken at 37°C for 20 minutes.

Results: The enzyme reaction produces high electron density cupric ferrocyanide, which deposits in mitochondrial membranes, cristae and crista spaces.

Note: Succinodehydrogenase is less tolerant to fixatives; thus, very low concentration aldehyde fixative is commonly used to fix for a short time. To prevent excessive enzyme inactivation, the sections can also be fixed with 6% hydroxy glyoxal for 30 minutes. It is better than glutaraldehyde in saving enzyme activities, whereas it is weaker in preserving cell microstructure.

15.1.3.3 Endoplasmic reticulum

More than 20 kinds of enzymes have been detected under electron microscopy, such as lactate dehydrogenase, peroxidase, esterase, thiamine focal phosphatase, glucose-6-phosphatase and nucleoside diphosphokinase. Glucose-6-phosphatase (G-6-P) has been regarded as marker enzyme of ER.

Compositions of incubation medium

0.2 mol/L Tris-maleate (pH 6.7)	20 mL
Distilled water	27 mL
G-6-P dipotassium salt	25 mg
2% lead nitrate	3 mL
Sucrose	4 g

Staining procedures
a. Fixation: G-6-P is less tolerant to fixative; hence, the glutaraldehyde fixation (immersed fixation) time should not more than 30 minutes. After fixation, the tissue is fully rinsed with Tris-maleate buffer (pH 6.7).
b. Sectioning: Routine presection, 30–40 μm in thickness.
c. Incubation: 30 minutes at 37 °C.
d. Postfixation: 1% cold osmium tetroxide for 30–60 minutes.

Results: G-6-P reaction products can be detected in cavity of ER, or in the intermembranous space of the nuclear membrane. When cells experience the degeneration stage, its G-6-P activity would drop significantly. For instance, in patients suffering from type I glycogen storage diseases (von Gierke disease) in which large amounts of glycogen are accumulated in hepatocytes, the G-6-P reaction is negative in ER, whereas in patients with type II glycogen storage diseases, its G-6-P reaction is positive in ER.

Notes
a. This method has strong staining effects but has slight diffusion phenomenon as well. To avoid the diffusions, reduce the amount of the substrates appropriately.
b. Under pH 5.0, the activities of the enzymes will be completely suppressed.
c. Hg^{2+} (1 mmol/L), NaF (20 mmol/L), Cu^{2+} (1 mmol/L), Zn^{2+} (10 mmol/L) and CN^- (10 mmol/L) are inhibitors of G-6-P. They can be used in the control experiment.

15.1.3.4 Golgi complex

The Golgi complex is associated with cell secretion. Thiamine pyrophosphatase enzymes (TPPase), nicotinamide adenine dinucleotide enzyme (NADPase) and cytidine monophospahatase (CMPase) are a set of Golgi complex marker enzymes, and osmiophilic reaction is unique to the Golgi complex.

Preparation for the incubation solution
a. Incubation solutions of lead citrate method showing TPPase

0.2 mol/L Tris-maleate buffer (pH 8.5)	1.4 mL
Distilled water	2.0 mL
Chloride TPPase	1.0 mL
0.015 mol/L magnesium sulfate	2.6 mL
0.5% high alkaline lead citrate solution pH 10.0	4 mL

b. NADPase incubation solution

NADP	1–2 mmol/L
Acetate buffer (pH 5.0)	40 mmol/L
Lead acetate	4 mol/L
5% Sucrose	146 mmol/L

c. CMPase incubation solution

C-5-MP	4 mmol/L
Acetate buffer (pH 5.0)	40 mmol/L
Lead acetate	4 mmol/L
Manganese chloride	5 mmol/L
4% Sucrose	117 mmol/L

Staining procedures

a. Fixation: Fix with formaldehyde calcium glutaraldehyde solution and completely rinse after fixation.
b. Sectioning: Presection in 30–40 μm thick slices.
c. Incubation: The lead citrate method of TPPase is the same as ALP enzyme incubation method. The lead acetate methods of NADPase, TPPase and CMPase are the same as the Gomori ACP enzyme incubation method.
d. Osmium tetroxide impregnation method: After sampling, the tissue block is immersed and fixed in 2% osmium tetroxide for 2 hours and then immersed in fresh 2% osmium tetroxide solution for a 48-hour impregnation at 37°C. Change the solution every 12 hours.

Results: Chemical reactions of the Golgi complex have certain specificity. The osmiophilic reaction zone locates at the forming face of the Golgi complex. NADPase activity is seen in the middle cisterna of the Golgi complex. The TPPase reaction zone is located at the mature face. CMP and ACP enzyme reactions appear in the Golgi-endoplasmic reticulum-lysosome region. At the stage of cell division, the saccules and vacuoles of the Golgi complexes disappear, and only the small vesicles are left. The change is difficult to indentify morphologically, and it has to be traced through marker enzymes of the Golgi complex.

15.2 Electron microscopic immunohistochemistry technology

Electron microscopic immunocytochemical technology, which is also known as immunoelectron microscopy technology, is the combined technology when the immunohistochemical staining results are examined under the electron microscope technology. It is a kind of method that studies the antigen-antibody combination and positioning in the ultrastructure level. It can be mainly divided into two kinds: one is immune agglutination electron microscope technology, that is, by means of antigen-antibody agglutination reaction, the products can be observed under an electron microscope after negative staining; another kind is immunoelectron microscopy positioning technology, that is, the specially labeled antibody combines with its corresponding antigen, and the complexes can be observed under an electron microscope. Because the markers that label the antibodies have certain degree of electron density, they can reveal the position of the corresponding antigens. The application of immunoelectron microscopy has brought the observation into the subcellular level.

15.2.1 Technical requirements of immunoelectron microscopy

In addition to the general requirements of immunohistochemistry, such as preparation, staining and observation, immunoelectron microscopy has some specific requirements.

15.2.1.1 Tissue preparation
Generally, the tissue samples should be fresh and exquisitely collected at low temperature as fast as possible, and all the solutions must be microporously filtered and prepared by buffer.

15.2.1.1.1 Fixation
The purpose of fixation is to preserve both the complete cell ultrastructures and the tissue antigenicity as much as possible. Among all the fixation methods, the perfusion fixation is recommended. The commonly used fixatives are paraformaldehyde-glutaraldehyde mixture and periodate-lysine-paraformaldehyde solution. The Bouin and the Zamboni solutions (paraformaldehyde and picric acid mixture) or 4% paraformaldehyde solution is also adopted.

15.2.1.1.2 Treatments after fixation
Long time processing in water-soluble reagent, dehydrating agent and liquid resin will cause loss of small molecule antigens and water-soluble antigens partially or completely, and cell ultrastructures will be damaged as well. These will finally result in negative staining. Therefore, the treatment time after fixation should be short as much as possible.

15.2.1.1.3 Embedding
a. Resin embedding: Currently, epoxy resin embedding is still the most used material. According to conventional procedures, a small piece of tissue is directly embedded after dehydration.

b. Low-temperature embedding: Because the conventional resin embedding requires high-temperature polymerization processing, tissue antigenicity may be lost completely or partially. Therefore, low-temperature embedding and frozen ultrathin section technology have been applied by many laboratories. The commonly used embedding media at low temperature are Lowicryls, LR White and LR Gold, and glycol methyl acrylate.

15.2.1.2 Immunostaining
The commonly used three types of methods include preembedded staining, postembedded staining and frozen ultrathin section embedded staining.

15.2.1.2.1 Preembedding staining
The slices are stained before ultrathin section. Fresh fixed tissues are presectioned into 40 μm thick sections by vibration microtome, succeeded by immunostaining (PAP method or others). After collecting the positive immunoreaction part from the tissue section under a dissecting microscope, the tissue is postfixed by osmium tetroxide, dehydrated, resin embedded, ultrathin sectioned and finally observed under EM.

15.2.1.2.2 Postembedding staining
Tissue samples are first fixed, dehydrated, resin embedded and cut into ultrathin section. Then the samples are mounted on the grid for immunostaining; thus, the method is also called on-grid staining. Because the ultrastructures of the tissues are preserved well in postembedded staining, and without the antibody penetrating problems, the continuous ultrathin sectioning or performing double (multiple) staining on the same ultrathin sections can be conducted. However, antigenicity may weaken in the process of samples preparation.

15.2.1.2.3 The frozen ultrathin section immunostaining
After the cells or tissues are mildly fixed, they are immersed in 2.3 mol/L sucrose (cryoprotectants) for 15–30 minutes or overnight to reduce the water content within tissues and to avoid the formation of ice crystals within tissue, which would affect the process of antigen localization. After the samples are completely soaked in sucrose, they are quick-frozen by liquid nitrogen and then sectioned by ultrathin microtome. Section thickness can be slightly thicker than that of regular resin section, and the sections will be finally mounted on nickel grid for immunostaining. Ultrathin freezing sectioning needs no procedures such as dehydration and embedding. It is immunostained directly; hence, antigen activities will be preserved well, and it has advantages of both preembedded and postembedded staining.

15.2.1.3 Markers
The immunoreactions are revealed by reaction products of high electron densities under an electron microscope. So far, the most commonly used markers are peroxidase and colloidal gold.

15.2.1.3.1 Peroxidase
Peroxidase is a stable small molecule. It is a common marker that easily penetrates tissue. It is commonly developed by a DAB chromogenic agent because the product of DAB is osmiophilic. After being oxidized with osmium, its electron density increases and can be easily detected under EM level. However, DAB reaction products tend to spread from *in situ* positions. Thus, 4-chloro-1-naphthol whose reaction products can also be oxidized with osmium and whose background is lower than that of DAB staining produces high specific staining after postembedded in immunoelectron microscopy and has been used as developing agents.

15.2.1.3.2 Colloidal gold
According to the magnification and labeling densities, in immunoelectron microscopic cytochemistry, the diameters of the gold particles should be carefully chosen, commonly 5–10 nm gold particles. Labeling density increases with the decreasing gold particle diameter. The smaller the gold particle diameter, the larger the total superficial area and the more IgG combined. However, if the gold particles are too small, it would not be easily detected under an electron microscope. Generally, the penetration ability of colloidal gold is weak. Thus, it is mainly used for postembedded staining.

15.2.1.4 Control test
To determine the specificity of methods, controlled tests are also required in immunoelectron microscopy technology.

15.2.2 Immunoenzyme electron microscopic technology

The principles of immunoenzyme electron microscopy are similar to those of light microscopy, except that it demands that the end products have higher electron densities. Preembedding staining and postembedding staining can both be used, and the commonly used method is the PAP method.

15.2.2.1 Preembedding staining

15.2.2.1.1 Staining procedures
Section preparation
a. Fixation: Perfusion fixation is frequently used.

b. Antifreeze protection: Tissue block is immersed into 30% sucrose solution for 6–8 hours or 4°C overnight, until tissue block sinks.
c. Freezing and thawing: Mend the tissue contour as small as possible (otherwise quick-freezing will cause tissue cracking), immerse it into liquid nitrogen for quick-freezing and then take out and put the tissue block into 0.1 mol/L PB, thawing at room temperature.
d. Sectioning in vibration microtome: The section thickness is approximately 40–60 μm. Store in PB at 4°C, and conduct the following treatments with floating sections.

Section pretreatments
a. Immerse the sections in 0.1% sodium borohydride and/or 20 mmol/L lysine for 30 minutes to remove the free aldehyde group.
b. Treat the section with 0.3% H_2O_2 role for 20 minutes to inhibit the endogenous enzyme activities.
c. If the tissue sections are not freeze-thaw treated, detergents such as 0.2% Triton X-100 or 0.01% saponins are suggested for 20 minutes, or the slices are treated with low concentration of ethanol to increase tissue penetrability.
d. Block the nonspecific binding sites in 3%–10% normal goat serum for 30 minutes.

Immunostaining
a. Immerse the sections in primary antibody, 4°C overnight or longer.
b. Rinse with Tris-buffered saline (TBS), 5–10 minutes, twice.
c. Immerse the sections in goat antirabbit IgG 1:100 at room temperature for 30 minutes.
d. Rinse with TBS as mentioned previously.
e. Immerse the sections in rabbit PAP complex 1:100 at room temperature for 30 minutes.

Note: The colloidal gold-labeled secondary antibody can also be used to take the place of the PAP method. It also gains satisfactory results.
f. Rinse with TBS as mentioned previously.
g. Immerse the sections in 0.05% DAB–0.01% H_2O_2 (0.1 mol/L, pH 7.6, TB preparation) at room temperature for approximately 5 minutes; control the staining time under light microscopic.
h. Rinse with TBS as mentioned previously.

Embedding method in electron microscopy
a. Fix in 1% glutaraldehyde for 10 minutes.
b. Rinse with TBS as mentioned previously.
c. Cut off positive reaction part under the dissecting microscope.
d. Immerse in 1%–2% OsO_4 for 1 hour; rinse with phosphate-buffered saline (PBS) two times.

e. Immerse in 50% ethanol uranium (100% ethanol: saturated aqueous solution of uranium acetate = 1:1) → 50% ethanol → 70% ethanol for 10 minutes in each solution.
f. Immerse in 95% ethanol, twice, and 100% ethanol, twice, for 15 minutes in each solution.
g. Treat with epoxy propane for 15 minutes, twice.
h. Resin: epoxy propane = 1:2, 30 minutes.
i. Resin: epoxy propane = 1:1, 30 minutes.
j. Change the resin twice, 30 minutes each time.
k. Plate embedding: Small pieces of tissue slices are embedded in the middle of two pieces of plastic covers; polymerize at 56°C for 24 hours.
l. Secondary embedding: Cut off the plastic cover, put it in a flat base capsule (cut off the tips of electron microscopy embedding capsule bottom and then invert upside down), inject resin and polymerize at 56°C for 48 hours.
m. Perform ultrathin sectioning, and observe under an electron microscope.

15.2.2.1.2 Note
a. Control tests are the same as in light microscopy.
b. Antibody diluents are prepared by TBS containing 1% normal goat serum.
c. If no positive reaction appears, verify the results by examining whether the above fixation and staining methods are positive in frozen sections. If not, remove glutaraldehyde from fixative and add 0.2% Triton X-100 in the fixative. If the results are positive, gradually add glutaraldehyde in the fixative until enough positive staining is obtained. If positive staining results still do not appear, it indicates problems of certain reagent. The most common reason lies in the primary antibody.

15.2.2.1.3 Results
The processes of immunoenzyme electron microscopy technology are complicated, and explanations of the results need to be with discretion (Fig. 15.1).

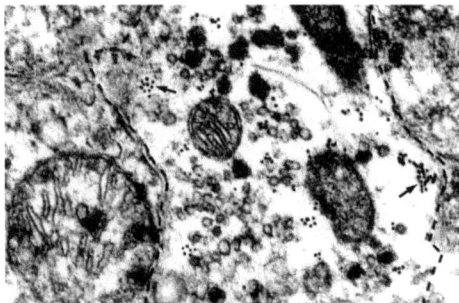

Fig. 15.1: Electron microscopic immunocytochemistry staining (colloidal gold-labeled secondary antibody method and preembedding staining). Arrows show colloid gold particles in hypothalamus axons of baboons.

15.2.2.2 Postembedding staining

15.2.2.2.1 Staining procedures
Electron microscopy embedding and sectioning: After routine perfusion fixation, sampling and electron microscopy embedding, semithin sections are picked up on slides, and ultrathin sections are loaded on nickel grid and are stained after drying.

The semithin section immunostaining

a. Immerse the sections in alcohol-NaOH (add NaOH in pure alcohol at 1:1 ratio, place for 1 week) solution for 5–10 minutes to remove resin.
b. Rinse the sections with 100% ethanol for 5–10 minutes.
c. Rinse the sections with water and then with TBS, twice.
d. Immerse the sections in 3% normal goat serum for 5 minutes.
e. Incubate the sections in the primary antibody for 4–6 hours at room temperature or 4°C overnight, and rinse with TBS, twice.
f. Immerse the sections in goat antirabbit IgG 1:100 for 30 minutes. Rinse with TBS, twice.
g. Immerse the sections in PAP 1:100 for 30 minutes. Rinse with TBS, twice.
h. Immerse the sections in DAB-H_2O_2 for 5 minutes. Rinse with tap water.
i. Perform dehydration, transparent and mounting as usual.

The ultrathin section on-grid staining

a. Etch the sections in 5% H_2O_2 solution for 5 minutes.
b. Rinse with TBS for several times, and filter paper dry from the nickel grid edges.
c. Immerse the sections in 3%–10% normal goat serum for 5 minutes, and paper blot up.
d. Immerse the sections in the primary antibody, 4°C overnight.
e. Rinse with TBS three times, each for 10 minutes, and blot up.
f. Immerse the sections in normal goat serum for 5 minutes.
g. Immerse the sections in goat antirabbit IgG 1:50–1:100 at room temperature for 5 minutes.
h. Rinse with TBS same as step e.
i. Immerse the sections in rabbit PAP 1:50–1:100 for 5 minutes, and rinse with TBS same as step e.
j. Immerse the sections in DAB-H_2O_2 for 3 minutes, and rinse with TBS same as step e.
k. Treat the sections with 4% OsO_4 for 20 minutes, rinse with DDW and blot up.
l. Observe under an electron microscope.

15.2.2.2.2 Note
a. Resin does not destroy the antigenic determinant, but it will delay and hamper the mobile of antibody. Etching (or other ways which could partially remove the resin) is helpful to expose antigenic determinants well. Therefore, the applications of semithin section have increased gradually.

b. Postembedded staining can be conducted on wax plate. Before staining, put a drop of reagent on the plate and then float the grid on the surface or in the droplets.
c. The rinsing in immunostaining must be thorough. Rinse methods include immersed rinse and spray rinse. In immersed rinse, the nickel grid is clamped and immersed repeatedly in beakers containing cleaning solutions. In spray rinse, the curved needles with plastic spray bottle or injection syringes are used as spray instruments. At the time of rinse, spray the cleaning solutions along with the surface of the nickel grid and ensure the sufficient solution volumes and spray speed are proper. Spray rinse is more suitable in order to achieve the purpose of rinse. Residual liquid on the grid can be blot up with filter paper.
d. Avoid of dries of nickel grid in the process of staining.

15.2.3 Immunogold electron microscopic technology

Because colloidal gold particles are large and it is difficult for them to penetrate tissue sections, the colloidal gold labeling technology is mainly used in postembedding staining. Preembedding staining can be used in the inspection of cell surface antigens and receptors, and it shows the intracellular antigens of single-layer cultured cells. In addition, the protein A gold technology (pAg) has strong specificity and light background staining. It has been more widely applied than colloid gold labeling technology in immunoelectron microscopy.

15.2.3.1 Postembedding protein A gold technology
Staining procedures
a. Embed the samples in resin, cut into ultrathin sections and load on nickel grid.
b. If the sample was not fixed by OsO_4, it does not need etching. Otherwise, the nickel grid should be put into 10% H_2O_2 for 10 minutes and then rinsed by PBS.
c. Float the nickel grid (with ultrathin section side down) in a drop of 1% ovalbumin-PBS buffer (pH 7.4) and keep for 5 minutes at room temperature.
d. Transfer the nickel grid (without flushing) into a drop of primary antibody for 2 hours at room temperature or 4°C overnight (in wet box). The dilution liquor of the primary antibody can be PBS or TBS buffer containing 1% ovalbumin, pH 7.4.
e. Spray rinse and immersion rinse for 2 minutes. Spray rinse again and blot up with filter paper.
f. Put the nickel grid on a drop of pAg solution for 1 hour at room temperature. The pAg is usually 10–20 times diluted.
g. Spray rinse with PBS for 1 minute, immerse in PBS for 5 minutes and then spray rinse and immersion rinse again.
h. Spray rinse with DDW and blot up with filter paper.

i. Immerse in 5% aqueous solution of uranyl acetate staining for 10–20 minutes, and rinse with water.
j. Lead citrate staining for 2–5 minutes, and rinse with water.
k. Observe under an electron microscope.

Note: If low-temperature embedding media are used, the counterstaining time should be shortened accordingly.
    Control test
a. Immunostain with antigen-absorbed primary antibody; the results should be negative.
b. Omit the primary antibody and use the pAg solution only; the results should be negative.
c. After incubation with the primary antibody during staining, unlabeled protein A solution (0.1–0.2 mg/mL) is first used to react for 1 hour, and then pAg solution is used; the results should be negative.

Use the normal nonimmune serum (the same animal species as that of first antibody) to replace the antiserum; the results should be negative. While in normal serum, Ig can combine non-specifically with the tissue and react with pAg, but the reaction degree is very low, and the background staining is less than that in absorption test.

15.2.3.2 Postembedding gold-labeled technology

a. The ultrathin sections are loaded in nickel grids.
b. Etch the sections in 10% H_2O_2 for 15 minutes.
c. Rinse the sections with DDW, three times, 10 minutes each time.
d. Nickel grids are immersed in or floated on 5% normal goat serum for 10 minutes.
e. Blot up the filter paper and incubate the sections in the primary antibody for 2 hours at room temperature or 4°C overnight.
f. Rinse with TBS for 5 minutes, three times.
g. Rinse with 0.1% bovine serum albumin (BSA)-TBS buffer (pH 8.2) for 5 minutes.
h. Immerse the sections in 1:30–100 gold-labeled secondary antibody for 10 minutes to 1 hour at room temperature.
i. Rinse with 0.1% BSA-TBS, three times.
j. Rinse with TBS for 5 minutes.
k. Postfix in 1% glutaraldehyde for 10 minutes.
l. Rinse with DDW for 5 minutes, three times.
m. Counterstain the sections with 5% uranyl acetate and lead citrate for 5 minutes, and rinse with DDW after each staining.
n. Observe the sections under an electron microscope.

15.2.3.3 Preembedding of single layer culture cells

a. Cells are cultured on the glass coverslips or plastic coverslips.
b. Rinse with PBS, two times, to remove the culture medium, and fix for 10 minutes in paraformaldehyde-glutaraldehyde (0.3%–1%). Triton X-100 (0.1%–0.1%) can be added in the fixative to increase penetrability if necessary.
c. Rinse with PBS rinse for 10 minutes, three times.
d. Immerse the cells in 1% sodium borohydride for 10 minutes.
e. Rinse with PBS for 10 minutes, three times.
f. Rinse with 0.1% BSA-TBS (pH 8.2) for 10 minutes.
g. Immerse the cells in 5% normal goat serum (prepared by adding 1% normal goat serum to 0.1% BSA-TBS buffer) for 20 minutes.
h. Immerse the cells in primary antibody (prepared by adding 1% normal goat serum to 0.1% BSA–TBS buffer) for 2 hours or 4°C overnight.
i. Rinse with BSA-TBS, three times.
j. Gold-labeled goat antirabbit IgG is appropriately diluted to stain the cells for 2 hours or 4°C overnight.
k. Rinse with BSA-TBS for 5 minutes, three times.
l. Rinse with PBS for 5 minutes, two times.
m. Postfix with 1% glutaraldehyde and 0.2% tannic acid for 30 hours.
n. Rinse with PBS for 5 minutes, three times.
o. Postfix with 0.5%–1% OsO_4 for 10 hours.
p. Rinse with PBS for 5 minutes, three times.
q. Sections undergo dehydrate and flat embed, polymerize at 50°C for 3 days.
r. Select the appropriate cells under phase contrast microscope, and cut the cells from the coverslip for the second embedding.
s. Ultrathin section and observe under an electron microscope.

15.2.4 The double immunostaining transmission electron microscope technology

15.2.4.1 Continuous ultrathin sectioning

In this method, the two adjacent continuous ultrathin sections are immunostained by the postembedded PAP method using different antibodies. This method will help in detecting the coexistence of antigens in the ultrastructure level. The IGS method could also be used.

15.2.4.2 Direct IGS double staining

The two kinds of primary antibodies are labeled by colloid gold of two different sizes, and then the two kinds of antigens can be displayed in one nickel grid (in one section) by using postembedded staining. This method is simple but less sensitive. The broad preparation of monoclonal antibodies has prompted the use of direct method, and the different cell surface antigens can be displayed by a one-step method or a two consecutive step method.

15.2.4.3 Indirect IGS double staining

15.2.4.3.1 Primary antibody derived from two different animal species
a. The resin-embedded ultrathin sections are etched in 10% H_2O_2 for 5 minutes and rinsed by DDW.
b. Immerse the sections in 3% normal goat serum for 10 minutes.
c. Incubate the sections in two kinds of primary antibodies mixture (from rabbits and guinea pigs) for 1 hour at room temperature.
d. Spray rinse with TBS buffer, 0.2% BSA-TBS and 1% BSA-TBS for 5 minutes.
e. Incubate the sections with 20 nm gold-labeled goat anti-guinea pig IgG and 12 nm (or 40 nm) gold-labeled goat antirabbit IgG mixture for 1 hour at room temperature.
f. Spray rinse with 1% BSA-TBS and DDW.
g. Counterstain with uranyl acetate and lead citrate, and observe under an electron microscope.

15.2.4.3.2 Primary antibody derived from the same animal species
The nickel grids are treated by heat formaldehyde vapor (80°C) for 1 hour between two IGS staining process. The treatment destroys the antibody binding sites in first immunostaining. The nickel grid is then rinsed, respectively, by DDW, TBS and BSA-TBS for 10 minutes, and then the second staining is conducted. Control test for this treatment is necessary.

15.2.4.4 Immunoenzyme and immunogold double staining
The combination of immunoenzyme and immunogold technology can be used together for preembedded double staining, but the process demands further antibody penetration treatments. Generally, the PAP method is first used to display one kind of antigen, and then the enhanced DAB method is used to show the antibody reaction sites in the first staining. The IGS method is then used to display another kind of antigen. Moreover, using the IGS method first and successively the ABC method or the PAP method is recommended as well. For instance, first, the rat hypothalamus thick sections are incubated with 1:1000 anti-tyrosine hydroxylase (TH) antibody overnight. Later on, the specimens are incubated with 5 nm gold-labeled goat antirabbit IgG for 2 hours, and the TH antibodies are displayed by silver enhancement processing. Second, 10% normal rabbit serum is used to block the unoccupied antigen sites. Finally, the specimens are incubated with glutamic acid decarboxylase antibodies, stained by the ABC method, developed by the DAB method, embedded after catalyzing with osmium, ultrathin sectioned and observed under an electron microscope (Fig. 15.2).

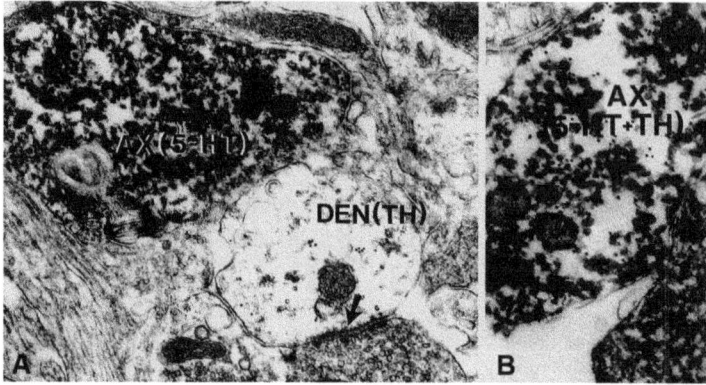

Fig. 15.2: Electron microscopic immunocytochemistry double staining method (immunoenzyme-immunogold method). (A) Baboon hypothalamus 5-HT immunoenzyme positive axons (AX) and tyrosine hydroxylase (TH) immunogold dendrites positive (DEN). (B) 5-HT and TH double positive axon.

In addition, the PAP method can be used as preembedded staining. Then the slices are embedded and postembedded stained by the IGS method.

15.2.4.5 Preembedding immunoenzyme double staining
Some electron donors, such as tetramethyl benzidine (TMB) and benzidine dihydrochloride (BDHC), are different in colors when compared with DAB. Thus, they can be used in light microscope double staining. Furthermore, the reaction products are easy to be distinguished because they have different shapes and electron densities. Therefore, these methods combined with the DAB method can be used in electron microscopic preembedded double staining

15.2.4.5.1 TMB
The preparation for chromogenic agents is as follows:

Liquid A: molybdenum acid or ammonium molybdate (0.24 g) is dissolved in 97.5 mL PB (0.1 mol/L), pH 6.0.

Liquid B: 5 mg TMB is dissolved in 2.5 mL absolute ethanol.

Before using, mix liquid A and liquid B, incubate the sections for 20 minutes, then add it in 0.3% H_2O_2 (1 mL) and incubate for another 15–20 minutes. Check the sections every 5 minutes until bluish-green color appears. Because the TMB reaction product color easily fades, before catalyzing with osmium, sections will be treated by DAB-cobalt stable liquid to stabilize the TMB reaction products. The preparation for stable liquid is as follows:

a.   Add 50 mg DAB in 100 mL of 0.1 mol/L TB, pH 7.4.

b.   Add 1% cobalt chloride (prepared by TB) 0.2 mL in former liquid.

c.   Check the reaction product until it becomes bluish-black, and then stop the reaction.

TMB reaction products are rodlike crystalline, 300–600 nm in length and 10–30 nm in diameter. They can also coagulate to form small pieces, which have high electron densities, and can be easily distinguished with DAB reaction products.

15.2.4.5.2 BDHC

Preparation for chromogenic agents: 10 mg BDHC is added in 95 mL DDW; stir to dissolve and filter. Add 25 mg nitro iron sodium cyanide and 0.2 mol/L PB (pH 6.8) 5 mL before using.

Incubate the section in former liquid for 10 minutes, and then incubate in chromogenic agents with 0.005% H_2O_2 (final concentration) for 5–10 minutes.

The color and the shape of BDHC reaction products are easily affected by pH value and ionic strength. Thus, before and after the incubation, the sections should be rinsed by 0.01 mol/L PB (pH 6.8, not more than 7.0) and oxidized with osmium prepared by the same buffer.

BDHC reaction products are dark-blue particles under the light microscope. Under electron microscopy, they are coarse particles aggregated by thin filaments or crystalline aggregation. They disperse in the cytoplasm and are different with DAB reaction products, which are cottonlike and diffuse distribution. In addition, their electron densities are much higher than that of DAB reaction products.

Note: In electron microscopy double staining, the PAP method and the ABC method are always used for the first staining, and the DAB method is used to display the first antigens. The developing time should be extended (for instance, 30 minutes) to ensure that a large number of DAB reaction products have been produced during the reaction. This will help cover up antigen binding sites and enzyme activities of the first chromogenic agents and prevent the cross reactions between two staining processes.

The second antigens are always displayed by TMB or BDHC, then osmium oxidized and embedded, ultrathin sectioned and observed under EM. However, this double staining method is not suitable for the two antigens, which exist in the same structural constitution or closely.

In addition, coarse reaction products of TMB or BDHC often shield and blur underneath detailed structures.

15.2.5 Immunostaining scanning electron microscope technology

Immunostaining scanning electron microscope (SEM) is a technology that uses SEM to study the three-dimensional structure of cell surface and the relationships between structure and surface antigens, thus locating antigens. In this technology, the sample preparation is simple, the three-dimensional distributions of markers are easy to observe and the observation area is large.

15.2.5.1 Markers

The markers used in SEM should be in the range of the SEM resolution and should have better capability of positioning the cell and the tissue antigens. Markers should be chosen based on the research targets.

Markers can be divided into three categories: particle markers, enzyme markers and radioactive markers. Particle markers are the most commonly used markers in SEM. According to the characteristics, these markers can be divided into the following: proteins such as hemocyanin and ferritin; pathogens such as tobacco mosaic virus, southern cowpea mosaic virus, bacteriophage T_4, *Escherichia coli* and phage; and metal particles such as colloidal gold, immunogold-silver technology and autoradiography silver particles. Among these, metal particles are the most widely used.

Colloidal gold is the most commonly used particle in SEM, and gold particles with 30–60 nm diameters are preferred. Moreover, because the gold itself is heavy metal and has strong ability to launch secondary electrons, it does not need to spray metallic membrane. Pathogen markers rely on their special appearance and structure to achieve the purpose of labeling and locating. For instance, bacteriophage T4 has a starlike racket, a hexagonal star-shaped head of 100 nm diameter and a tail of approximately 100 nm length; tobacco mosaic virus is a 15 × 30-nm rod-shaped virus; southern cowpea mosaic virus is a spherical particle with a diameter of 25 nm; and ferritin has a dense ferric ion core, and the high electron density of ferric ion can be used in positioning.

15.2.5.2 Immunolabeling method

Metal markers in immunolabeling are the same as in immunostaining; that is, the markers combine with antibodies, and then the antigen sites are displayed through direct or indirect methods. The colloidal gold can combine with protein A, followed by the combination of protein A and Fc section of IgG molecules. The latter is able to react with biotin-labeled antibodies. In immunogold-silver staining method, the targets are first labeled by colloidal gold and developed by silver fluid.

Virus markers are not usually labeled (unlabeled antibody method, or known as the bridge method). The theory of which is to use the same animal species to prepare the antigen-specific antibodies and virus marker-specific antibodies, respectively (e.g., rabbit anti-A antigen and rabbit anti-HRP antibodies). Then another animal species is used to produce the bridge antibodies (e.g., goat antirabbit IgG antibodies). In the process, the goat antirabbit IgG is used as a bridge to combine with the antigen-specific antibodies and the virus marker-specific antibodies, thus achieving the purpose of locating the antigens. The theory of which is similar to the PAP method, but the unstained virus usually shows the negative special appearance under SEM.

Pathogen immunomarkers can be displayed without markers; it can be displayed and positioned by using its morphological features or antigen-antibody agglutination method. Its basic theory is to use antigen-antibody reactions between virus or virus antigen-specific antibodies and their corresponding antigens. The reactions can cause the crosslinking of antigens and the successive agglutination. The concentration of the latter could be displayed under an electron microscope after negative staining.

Knowledge links: Total internal reflection fluorescence microscopy

With the developments of life sciences, scientists need to understand the relationship between chemicals and how they influence the cell functions at the levels of single cell and molecule. The invention of the total internal reflection fluorescence micro-scopy (TIRFM) satisfies some of the needs mentioned previously.

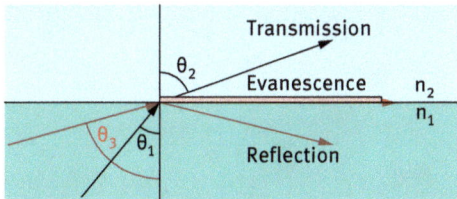

Fig. 15.3: Theory of total internal reflection fluorescence microscopy.

Total internal reflection is one of the natural phenomena. When the incident light from medium 1 with refractive index n_1 and incident angle θ_1 comes to another medium 2 with refractive index n_2, the light will be divided into transmission light with transmission angle θ_2 and reflection light. In this case, the following formula can be applied: $n_1 \sin\theta_1 = n_2 \sin\theta_2$. If $n_1 > n_2$, for example, the incident light comes from glass and goes to a solution, and if θ_1 reaches its critical value, the θ_3, no transmission light exists. This phenomenon is known as total internal reflection (Fig. 15.3). When total internal reflection occurs, because of the undulation effect, some of the light power could still go into medium 2, which is known as the evanescent field. The eva-nescent field is an inhomogeneous wave. It runs along the interface of medium 1 and medium 2. However, its amplitude of vibration exponentially decays at vertical inter-face, which means the evanescent field only exists at the interface and its thickness is approximately one wavelength, and it will be several hundreds of nanometers for visible light. In this way, the evanescent field can be used to stimulate the fluorescent chemicals that exist very close to medium n1.

Most of the cell activities, such as signal transduction, protein transportation, myosin movement, ATPase turnover, cell endocytosis and exocytose and pathogen invasion, are the events at molecular level. If marked by proper fluorescence, TIRFM will play an important role in those studies.

16 Quantitative assay of histochemistry experiment results

In the past recent decades, many kinds of methods and instruments for quantitative assay were created based on the fast developing physics theories and technologies, which make it possible to save and analyze the histochemistry results at a quantitative level easier, prompt and accurate. Of all the creations, the most used are photomicrography, image analysis, flow cytometry and laser scanning confocal microscopy, etc.

16.1 Photomicrography

Unlike the regular camera, photomicrography needs to take pictures of tiny structures at least in the micrometer level; thus, the apparatus and the picture-taking technology are different.

16.1.1 Basic theory

The charge-coupled device (CCD) is the mostly used technology for taking and recording pictures in photomicrography, and the pictures can be transferred, observed and saved in a computer.

16.1.1.1 Basic apparatus in photomicrography

The basic apparatus in photomicrography includes a common light microscopy, a digital camera and a computer that has a specific software and is connected to camera. Unlike the common camera, this digital camera has no lens and is installed on the drawtube of the light microscopy (Fig. 16.1). After the observation field is selected, the specific software is used to take and save the pictures.

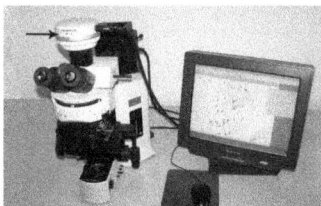

Fig. 16.1: Photomicrography. The arrow shows CCD.

DOI 10.1515/9783110531398-016

16.1.1.2 CCD

As an image sensor, the rectangle CCD is the crucial part of a digital camera. CCD is a photosensitive semiconductor chip and is composed of millions of tiny photodiodes that function as film in the regular camera. Each photodiode is sensitive to weak light and generates an electric charge, which undergoes a digital-to-analog conversion, forming a pixel. Most cameras use the red, green and blue monochromatic filter, and the color image is formed by additive color process.

16.1.1.3 White balance

White balance is one of the easy-to-ignore adjustment functions before taking the picture. The function of white balance is to make sure the white background under a microscope still appears as a white color in picture. Before taking the pictures, making a white balance is very important, along with focusing and choosing the right exposure time and film speed (Fig. 16.2).

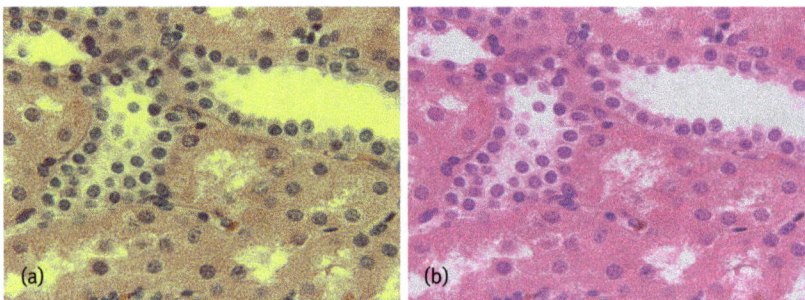

Fig. 16.2: Before (a) and after (b) white balance.

16.1.2 Application

The purpose of histochemistry staining is to produce the distinguished deposits *in situ*, which can be analyzed quantitatively and qualitatively. Thus, how to take and save the pictures under light microscopy for future analysis, discussion and publication becomes a basic need for histochemistry. In this way, the photomicrography is widely used.

16.2 Image analysis

The image analyzer is one of the modern precise instruments for morphology assay. The image analyzer used for biomedical research mainly includes microscope, image input unit (camera or scanner), special computer and information output unit (cathode-ray tube screen or printer). The special computer collects and analyzes the images, and the information of morphological quantity, the gray scale and even the three-dimensionality of the pictures are acquired. Thus, the image analyzer is a system to analyze the histochemistry pictures.

16.2.1 Theories of image analyzer

The image lights are reflected on the camera screen, which will be transferred into strong and light electric charges by electronic instruments and then undergo the digital-to-analog conversion and finally redisplayed on the screen for observation and analysis.

16.2.2 Working procedures of image analyzer

16.2.2.1 Collection and management of pictures

The pictures taken by camera and transferred into the computer must be saved as digitalized information, and digitization processes need to take the gray-scale measure from samples in the pictures to form a pixel. The entire procedure needs a perfect camera and is processed in the computer.

16.2.2.2 Image assay

With the gray-scale criterion, which needs to be set up before the assay, the image analyzer will calculate the corresponding parts of the pictures. If the color image analyzer is used, different colors can also be set up as the criterion.

16.2.2.3 Image analysis

The picture on the screen is composed of numerous dots. The more dots per square, the clearer the image. Each dot owns two sorts of information: the gray scale and the location in the picture. These sorts of information determine the shape and the density of the picture.

16.2.2.3.1 Assay of morphological parameter

The entire assay is conducted based on the pixels. When the need-to-know field or cell is selected on purpose with a computer mouse, the geometric parameter, such as perimeter, diameter (the largest and smallest), area and shape, will be measured by computer analysis.

16.2.2.3.2 Gray scale
The gray scale refers to the gradation of light, which can be divided into 256 levels by image analyzer: 0 means pure black, and 256 means pure white. The image analyzer usually gives dozens of gray scales for one picture, some of which are undistinguished by naked eyes.

In the color image analyzer, the different colors can be designated into different gray scales. In this case, the true color will be reflected on the screen for a colorful observation, and the fake color will enhance the contrast ratio of the picture for the analysis.

16.2.2.3.3 Measurement of optical density
The more advanced image analyzer houses the spectrodensitometer, which will digitally and logarithmically transfer the image information that will be modified based on absolute optical density. Thus, the spectrodensitometer mainly detects and outputs the integrated optical density, which is the total optical density distribution of every detected pixel. Because the area of objective can be easily calculated, the average optical density will be easily known. Unlike the gray scale, which is the relative value, the optical density is the absolute value and is directly proportional to the shade of the image.

16.2.3 Application of image analyzer in biomedical research

The image analyzer develops very fast in recent biomedical research because not only the geometric parameter of cells and characteristic parameters of the tissues but also the information of two-dimensional and three-dimensional reconstruction can be acquired. The need to know the size, square and volume of tiny structures under microscope, the advantages of image analysis and other newly created instruments for measurement gradually bring out a new discipline – stereology.

16.3 Flow cytometry

As a newly applied technology in the recent 20 years, the precise quantitative analysis and classification for a single cell with a speed of more than 10 million cells per second can be conducted by flow cytometry (FCM). The parameters of nucleus acid, the cell volume, the nucleus-to-cell ratio and the immune-marked proteins can be detected. Under the sterile circumstances, the cells can be collected by classifications, and its purity reaches 90%–99%. This high-speed, multi-information and high-purity technology provides the cell cycle kinetics research with a powerful method, and it is more and more popular in immunobiology, somatic cell genetics, radiation biology and oncology research.

16.3.1 Main structure and basic theory of FCM

FCM is mainly composed of fluidic system, exciting light device, signal measurement system and control system. Some kinds of FCM have a cell-sorting function with a corresponding device. The basic theory of FCM is that the florescence-marked single-cell suspension is prepared and enters into the sample tube and then the flow chamber by air pressure. The sheath fluid in the flow chamber restrains the cells in the sample to be arranged in a single line, forming a cell liquid column, which will be detected by perpendicular exposure to exciting laser light beam in the measurement area (Fig. 16.3). When exposed to exciting light, the emission light from the fluorescence in the cell will be detected, and the cell volume or other information such as the content of special materials in the cell can be examined. In addition, after the sample is highly diluted and passes through the exposure area, FCM will immediately determine whether the cell is useful or useless for further studies according to the fluorescence marker, and an electric charge is added to the single droplet containing one single cell. The charged and uncharged droplets will go through different pathways under the influence of magnetic field generated by charged deflection plates; thus, the marked and unmarked cells are sorted.

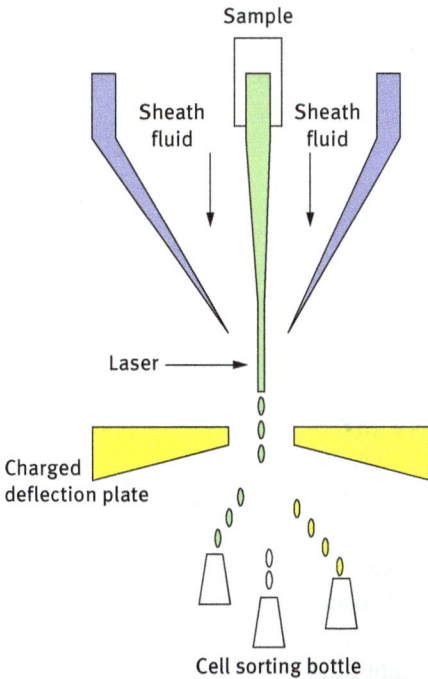

Fig. 16.3: Theory of FCM.

16.3.2 Sample preparation for FCM

The sample preparation is crucial for FCM.

16.3.2.1 Cells in culture
a.   The culture medium is discarded when cells are in their logarithmic growth phase.
b.   A 0.25% pancreatic enzyme is added to rinse the cells, and then the same 1–2 mL enzyme is added to treat the sample for 2–3 minutes. When cells are ready to shed, the enzyme solution is discarded by centrifugation at 1500 rpm for 5 minutes.
c.   The 3–4 mL non-Ca^{2+} or Mg^{2+} phosphate-buffered saline (PBS) is added to prepare the cell suspension with the pipette beating method, and the suspension is moved into the centrifugation tube.
d.   The suspension undergoes centrifugation at 1500 rpm for 5 minutes, and then the deposits and 0.5 mL liquor are collected.
e.   The 4°C 70% alcohol is added to fix the cells with the pipette beating method, and the sample can be kept at 4°C for 2 weeks.

16.3.2.2 Fresh abdominal dropsy and solid tumor
Fresh abdominal dropsy
a.   PBS (3 mL, pH 7.4) is added into 0.5–1 mL fresh abdominal dropsy. PBS is composed of the following chemicals:

NaCl	8.0 g
KCl	0.2 g
$Na_2HPO_4 \cdot 12H_2O$	2.9 g
KH_2PO_4	0.2 g
H_2O	1000 mL

b.   It is centrifuged at 1500 rpm for 5 minutes and rinsed with PBS, twice. The supernatant fluid is discarded.
c.   The 4°C 70% alcohol is added to fix the cells with the pipette beating method, and the sample can be kept at 4°C for 2 weeks.

Solid tumor
a.   The solid tumor is digested with 0.25% pancreatic enzymes.
b.   It is cut into pieces with scissors and undergoes centrifugation as previously mentioned.
c.   It is rinsed with PBS twice, and the supernatant fluid is discarded.
d.   The 4°C 70% alcohol is added to fix the cells with the pipette beating method, and the sample can be kept at 4°C.

16.3.2.3 Paraffin-embedded tissue

a. The paraffin-embedded tissue is sectioned into 40–50 µm thick slides, and three to five sections are collected into the tube.
b. Xylene (3 mL) is added to dewax the tissue for 1–2 days.
c. Alcohol (5 mL) is successively added at 100%, 95%, 70% and 50%, each for 10 minutes.
d. Alcohol is removed, and 3–5 mL double-distilled H_2O is used to rinse the tissue for 3–5 minutes.
e. Pepsin (5 mL, 0.5%), pH 1.5, is added.
f. Digestion undergoes 30 minutes at 37°C and is interrupted by 5 mL normal saline.
g. A 300-mesh nylon is used to filter the sample. Then the liquid undergoes centrifugation, and the supernatant fluid is discarded.
h. The sample is rinsed with normal saline twice, and the supernatant fluid is discarded.
i. The 4°C 70% alcohol is added to fix the cells with the pipette beating method, and the sample can be kept at 4°C.

16.3.2.4 Staining method

The parameters detected by FCM in the cells must be marked by fluorescence, which could be ethidium bromide (EB) or propidium iodide (PI), which stains cell nucleus. The fluorescence of PI is approximately 1.8 times stronger than that of EB, and PI will insert into double strands of DNA. In addition, mithramycin and Hoechst 33258 will specifically combine with G-C and A-T, respectively. 4.6-Diamidino-2-phenylindole hydrochloride will attach to the nucleus, and acridine orange (AO) will attach to DNA and RNA, showing different colors. As for the proteins, fluorescein isothiocyanate is the most used fluorescence.

16.3.2.4.1 DNA staining with PI

a. The cell suspension stored in 70% alcohol undergoes centrifugation at 2000 rpm for 5 minutes, and alcohol is discarded.
b. Rinse with PBS twice and centrifuge as above.
c. The cell concentration is adjusted by PBS to 1×10^6/mL.
d. RNase (0.5 mL, 0.5%) is used to digest at 37°C for 30 minutes.
e. PI (1.5 mL, 0.05%) is introduced to stain DNA at 37°C, at least for 30 minutes.
f. The suspension is filtered by 300-mesh nylon meshes, and 10^4 cells are detected by FCM with excitation light wave at 488 nm.

16.3.2.4.2 DNA-RNA double staining with AO

AO stock solution preparation

AO	50 mg
DDW	50 mL

Stored at 4°C for 4 months.

The AO stock solution is diluted at 1:10 by double-distilled water and then by Tris-HCl buffer (50 mmol/L Tris, 25 mmol/L KCl and 25 mmol/L $MgCl_2$) to a final concentration of 8.5 μg/mL.

Staining

a. The 10^5 cells are collected, and the fixative is rinsed.
b. A 2-mL AO work solution is added to stain the sample for 10 minutes at room temperature and detected by FCM.

16.3.3 Application of FCM

16.3.3.1 Application in oncology

Almost all the tumors contain the stable multiploid DNA at a particular site or time, and most of them are aneuploid. The changes of DNA content can be used as evidence to guide the tumor treatment, to understand the tumor development and to judge the prognosis.

16.3.3.1.1 Tumor diagnosis analysis

FCM has been used to analyze the DNA ploidy in animals and in human beings. Almost all the tumor cells contain the DNA stem ploidy; thus, FCM can be used to diagnose the sample such as urinary bladder rinse fluid, chest and abdomen transudate, and sputum and bronchial lavage fluid, which are difficult to analyze in routine procedures.

16.3.3.1.2 Judgment of tumor prognosis

FCM analysis for nucleus acid is an important indicator of tumor treatment prognosis. The tumor that contains aneuploid or multiploid DNA or in high-rate S-phage DNA usually will suggest a poor prognosis, and vice versa.

16.3.3.1.3 Antitumor drug screening

The high efficient antitumor drugs not only kill the cancer cells in different phages of cell cycle but also function on a dot in cell cycle to delay cancer cell development to slow down cell proliferation, which enhances or decreases the function of chemotherapy drug. FCM plays an important role in drug screening and makes it possible for the antitumor drugs to reach their best effects.

16.3.3.1.4 Analysis of chromosome karyotype

The results of DNA content analysis by FCM can be shown in the image of chromosome frequency distribution. The same type of chromosome appears in one peak, and the area of the peak stands for their concentration. This method can quickly analyze the chromosome type and collect the same type chromosomes for further research.

16.3.3.2 Application in immunology

The parameters of immunocompetent cells can be detected, and the cells can be classified by FCM for future culture to process the research, for example, the characteristics of antibody combination in infected and uninfected cells and the cell cycle changes after infection. In addition, because FCM can highly and accurately process detection, the functional mechanism of immune disease can be evaluated.

16.3.3.3 Application in apoptosis

Apoptosis is another pathway of cell death. During apoptosis, the chromosomes break into pieces, the cell nucleus shrinks and the cell becomes densely stained and forms the apoptotic bodies that are engulfed by adjacent cells. It is demonstrated that apoptosis is different from cell death because the cell still synthesizes DNA, protein and ATP during apoptosis; that is, apoptosis is the initiative process of cell suicide. Another feature of apoptosis is that the DNA undergoes segmentation and shows a ladder after gel electrophoresis. FCM analysis shows that before the DNA G0/G1 peak, there is a hypodiploid (also known as hypo-G1) peak that stands for apoptosis, and the area beneath this peak stands for the relative number of apoptotic cells that can be measured by FCM (Fig. 16.4).

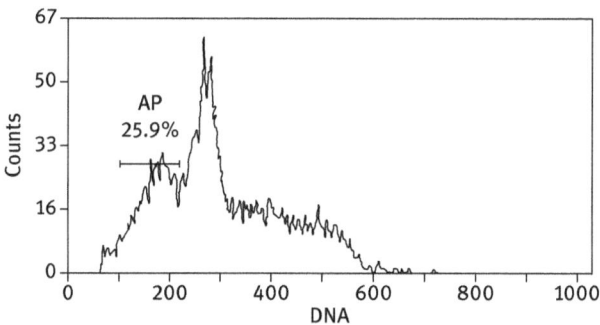

Fig. 16.4: FCM results for apoptosis examination. Count stands for the relative number of cells. The hypo-G1 peak is obvious (25.9%), which means apoptosis. AP, apoptotic peak.

16.4 Laser scanning confocal microscopy

The laser scanning confocal microscopy (LSCM) appeared in the 1980s. It is a highly advanced method to analyze the tissues and cells. In this magic technology, laser technology and computer science are combined together to cut the tissues and cells into the "light section", which will be examined by microscope, and the results will be analyzed and stored in the computer. In this way, the morphological quantitative observation, the physiological dynamic detection and the biochemical quantitative analysis are integrated. Because of the high sensitivity and the ability to detect spatial spots freely, LSCM could be observed from the superficial, single-layer, regional and static status to the three-dimensional, Tomoscan, dynamic and all-sided structures. With these advantages, the LSCM is rapidly introduced in the life science studies.

16.4.1 Theory and characteristic of LSCM

16.4.1.1 Theory of LSCM

The LSCM uses laser as light source, whose light wavelength is 351–364, 422, 488, 514 and 633 nm. The laser reaches and focuses on the sample via incident pinhole to form a light spot, whose reflection undergoes confocal and reaches imaging projection plane via observation pinhole. Thus, the reflection from nonconfocal plane and other background lights will be shielded (Fig. 16.5). In this way, the resolution of LSCM is enhanced. When LSCM scans along the X and Y axes and at the meantime the laser can also be adjusted to focus along the Z axis, the light section is created. The information from the X, Y and Z axes is collected by the computer. The XY, XZ, YZ and other arbitrary planes and the three-dimensional images can be observed. Because the rapid scanning is controlled by the computer, it is convenient for LSCM to detect the cell dynamic statue to understand the physiological and biochemical processes. In addition, the morphological and quantitative analyses of biological sample and the cell surgery (such as chromosome incision and application of laser tweezers) can also be processed by LSCM.

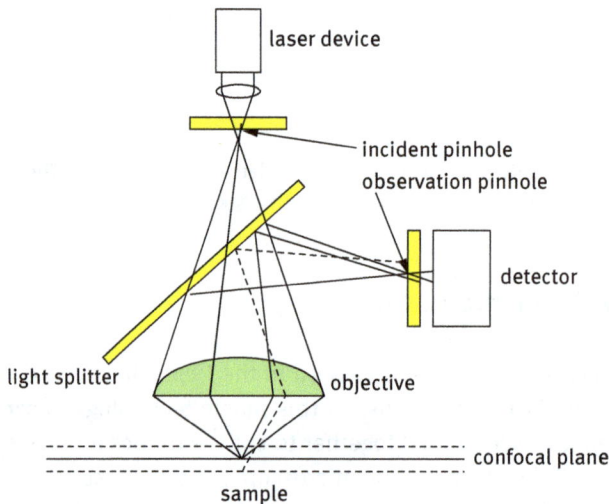

Fig. 16.5: Theory of LSCM.

16.4.1.2 Characteristic of LSCM

16.4.1.2.1 High resolution
The LSCM uses laser as light source to detect the sample, which will form an image with higher contrast than that in normal microscope. In addition, the sample information comes from the same focal plane, which means that the resolution is enhanced. According to the numerical aperture of objective lens, the minimum thickness of the sample is 0.5 μm, and the resolution is 1.4 times higher than that in normal microscope.

16.4.1.2.2 High sensitivity
In the LSCM, the fluorescence is detected by photomultiplier, forming a fluorescent digital image. The different light filters separate different fluorescence, and the very weak fluorescence can be detected.

16.4.1.2.3 High scanning speed
The LSCM can scan the sample at the microsecond level; thus, it can be used to study the changes of calcium ions, magnesium ions and pH values in alive cells and to determine membrane potential and mobility.

16.4.1.2.4 Light section and cell computed tomography
The LSCM uses laser to scan the sample dot by dot at a certain depth to create an XY axis plane image, and it can also scan the sample along the Z axis dot by dot or line by line. When the plane of fracture of the cell is studied, the computed tomography of the cell is being created.

16.4.1.2.5 Convenience for image access
Image data are collected by CCD and stored in the computer, easy for saving and analyzing the image.

16.4.2 Function of LSCM

16.4.2.1 Static status observation

16.4.2.1.1 Observation and analysis of sample with simple stain
The single fluorescence is used to stain the sample. The LSCM can detect and collect the data of average and integral fluorescence density, the cell perimeter and section area of the cells and other morphological factors of the cells, and the data can be saved as a column diagram.

16.4.2.1.2 Observation and analysis of sample with double stain

The two different fluorescence methods are used to stain the sample. The emission light wave of the fluorescence must be different to avoid the interruption of observation. These images will undergo overlaying by using a special software in the computer.

16.4.2.1.3 Quantitative assay of fluorescence density

This assay can accurately monitor the cell conjugation to try to understand the mechanism of autoimmune disease, AIDS, the function of drugs on diseases, the cytotoxicity experiments of combination of NK cells and target cells and the expression of antigens in different kinds of tumor sections.

16.4.2.2 Dynamic status observation

16.4.2.2.1 Assay of concentration of different kinds of ions in cells

Ca^{2+}, Mg^{2+} and pH value can be detected by LSCM at microsecond and millisecond levels by spot, line and square scanning.

16.4.2.2.2 Assay of mobility of cell membrane

After the cell membrane fluorescence probes are excited by polarized laser, the polarization of the emission lights depends on the rotation of the fluorescence molecules, and these well-organized free movements rely on the cell membrane mobility just around the fluorescence molecules. In this way, the LSCM indirectly reflects the mobility of cell membrane. The assay of cell membrane mobility plays important roles in the analysis of the composition of membrane lysophosphatidic acid, the effects and acting sites of drugs, the determination of temperature reaction and the comparison of different species.

16.4.2.3 Studies on cell communication

Cell communication mediated by gap junction plays an important role in cell proliferation and differentiation, and this oncogene- or chemical-modified gap junction could be a start signal for tumor formation. LSCM can be used to assay cell communication via fluorescence recovery after photobleach (FRAP) technology. In the FRAP technology, the high-intensity pulsed laser irradiation quenches the fluorescence in the exposure area. Meanwhile, the unquenched fluorescence will move to the exposed area via gap junction, and this process can be detected, observed and recorded by LSCM. Fluorescence 5-carboxyfluoresceindiacetate-aeetoxymethyleste (CFDA-AM) is the mostly used marker in FRAP technology.

16.4.2.4 Assay of caged and uncaged bioactive chemicals

The activity of bioactive chemicals (such as second messenger, nucleotide, neurotransmitter and Ca^{2+}) can be inhibited when combined with the caged compound, forming a caged status. When exposed to the instant laser of the specific wavelength from LSCM, the bioactive chemicals are released from the compound and begin to function. Because the wavelength, density and time of the exposure are under the control, the bioactive chemicals will function regionally, which can be detected by LSCM.

16.4.2.5 Selection of adherent cells

The adherent cells in culture cannot be selected by FCM, but LSCM owns the ability to select the adherent cells without damage to culture circumstance and cell growth status. The cells are cultured on special dish, and the high-power laser chooses the area containing the objective cells and cut the area in octagon, and the other cells outside the octagon will be removed. This method is suitable for the selection of minority cells such as mutant cells, hybrid cells and transferred cells. Another way to select the cells is that the unnecessary cells are marked by fluorescence and killed by high-power laser, and the remaining cells are kept and cultured for further studies.

Review question

What are the structures and functions of other microscopes, such as ultra high-resolution light microscopy and atomic force microscope?

Recommended readings

[1] Jicheng Li. Experimental Technologies of Histology and Embryology. Beijing, China, People's Medical Publishing House 2010.

[2] He Li, Li Zhou. The Technologies of Histochemistry and Cytochemistry. Beijing, China, People's Medical Publishing House 2014.

[3] Freida L C, Christa H. Histotechnology: A Self-instructional Text, 3rd edition. Chicago, USA, American Society for Clinical Pathology Press 2009.

[4] Richard W B. Immunocytochemistry. New York, USA, Springer Press 2009.

[5] Jinsong Zhou. The Modern Technology and Method of Histochemistry. Xi'an, Xi'an Jiaotong University Publishing House 2015.

[6] Xiaoqing Zhang, Shangwenyuan Gong, Yu Zhang. Prussian blue modified iron oxide magnetic nanoparticles and their high peroxidase-like activity. J Mater Chem 2010, 20, 5110–6.

[7] Dewar R, Fadare O, Gilmore H, et al. Best practices in diagnostic immunohistochemistry: myoepithelial markers in breast pathology. Arch Pathol Lab Med 2011, 135(4):422–9.

[8] Juniku-Shkololli A, Manxhuka-Kerliu S, Ahmetaj H, Khare V, Zekaj S. Expression of immunohistochemical markers of progression in pre-cancerous and cancerous human colon: correlation with serum vitamin D levels. Anticancer Res 2015, 35(3):1513–20.

[9] Gu J, Lei Y, Huang Y, et al. Fab fragment glycosylated IgG may play a central role in placental immune evasion. Hum Reprod 2015, 30(2):380–91.

DOI 10.1515/9783110531398-017

Appendix 1 Commonly used buffer in histochemistry

I Phosphate buffer, pH 5.2–6.8 (Tab. 1)

Liquid A: 0.1 mol/L KH_2PO_4
KH_2PO_4 1.361 g, add distilled water to 100 mL
Liquid B: 0.1 mol/L $Na_2HPO_4 \cdot 2H_2O$
$Na_2HPO_4 \cdot 2H_2O$ 1.78 g or $Na_2HPO_4 \cdot 12H_2O$ 3.581 g, add distilled water to 100 mL

Tab. 1: Phosphate buffer, pH 5.2–6.8.

pH	liquid A/ml	liquid B/ml	pH	liquid A/ml	liquid B/ml
5.3	9.75	0.25	6.81	5.0	5.0
5.6	9.5	0.5	6.98	4.0	6.0
5.91	9.0	1.0	7.17	3.0	7.0
6.24	8.0	2.0	7.38	2.0	8.0
6.47	7.0	3.0	7.73	1.0	9.0
6.64	6.0	4.0	8.04	0.5	9.5

II Tris-HCl buffer (0.05 mol/L), pH 7.19–9.10 (Tab. 2)

Liquid A: 0.2 mol/L Tris
Tris 2.423 g, add distilled water to 100 mL
Liquid B: 0.1 mol/L HCl
HCl 0.84 mL, add distilled water to 100 mL

Tab. 2: Tris-HCl buffer, 0.05 mol/L, pH 7.19–9.10.

pH	liquid A/ml	liquid B/ml	DDW/ml	pH	liquid A/ml	liquid B/ml	DDW/ml
7.19	10	18	12	8.23	10	9	21
7.36	10	17	13	8.32	10	8	22
7.54	10	16	14	8.41	10	7	23
7.66	10	15	15	8.51	10	6	24
7.77	10	14	16	8.62	10	5	25
7.87	10	13	17	8.74	10	4	26
7.96	10	12	18	8.92	10	3	27
8.05	10	11	19	9.10	10	2	28
8.14	10	10	20				

DOI 10.1515/9783110531398-018

III Acetic acid buffer, pH 3.6–5.6 (Tab. 3)

Liquid A: 0.1 mol/L acetic acid
Liquid B: 0.1 mol/L sodium acetate

Tab. 3: Acetic acid buffer, pH 3.6–5.6.

pH	liquid A/ml	liquid B/ml
3.6	185	15
3.8	176	24
4.0	164	36
4.2	147	53
4.4	126	74
4.6	102	98
4.8	80	120
5.0	59	141
5.2	42	158
5.4	29	171
5.6	19	181

IV Dimethylarsinic acid buffer, pH 5.0–7.4 (Tab. 4)

Liquid A: 0.2 mol/L sodium dimethylarsinate
Sodium dimethylarsinate 4.28 g, add distilled water to 100 mL
Liquid B: 0.2 mol/L HCl
HCl 1.7 mL, add distilled water to 100 mL

Tab. 4: Dimethylarsinic acid buffer, pH 5.0–7.4.

pH	liquid A/ml	liquid B/ml	DDW/ml	pH	liquid A/ml	liquid B/ml	DDW/ml
5.0	25	23.5	51.5	6.4	25	9.2	65.8
5.2	25	22.5	52.5	6.6	25	6.7	68.3
5.4	25	21.5	53.5	6.8	25	4.7	70.3
5.6	25	19.6	55.4	7.0	25	3.2	71.8
5.8	25	17.4	57.6	7.2	25	2.1	72.9
6.0	25	14.8	60.3	7.4	25	1.4	73.6
6.2	25	11.9	63.1				

V Citric acid buffer, pH 3.0–6.2 (Tab. 5)

Liquid A: 0.1 mol/L citric acid
Citric acid anhydrous 1.921 g, add distilled water to 100 mL
Liquid B: 0.1 mol/L citrate sodium
Citrate sodium (containing 1 H_2O) 2.941 g, add distilled water to 100 mL

Tab. 5: Citric acid buffer, pH 3.0–6.2.

pH	liquid A/ml	liquid B/ml	pH	liquid A/ml	liquid B/ml
3.0	46.5	3.5	4.8	23.0	27.0
3.2	43.7	6.3	5.0	20.5	29.5
3.4	40.0	10.0	5.2	18.0	32.0
3.6	37.0	13.0	5.4	16.0	34.0
3.8	35.0	15.0	5.6	13.7	36.3
4.0	33.0	17.0	5.8	11.8	38.2
4.2	31.5	18.5	6.0	9.5	41.5
4.4	28.0	22.0	6.2	7.2	42.8
4.6	25.5	24.5			

Appendix 2 Histochemistry and immunohistochemistry experiments

Experiment 1: Periodic acid Schiff (PAS) staining

PAS staining is mainly used for the detection of structures containing a high proportion of carbohydrate macromolecules (glycogen, glycoprotein and proteoglycans) in tissues such as liver, cardiac and skeletal on formalin-fixed, paraffin-embedded tissue sections, and it may be used for frozen sections as well.

I Materials and reagents

1. Periodic acid solution (1%)
Periodic acid	1 g
Deionized or distilled water	100 mL
2. Schiff reagent
Basic fuchsin	1 g
Distilled water	100 mL
Potassium (sodium) metabisulfite	2 g
1 mol/L hydrochloric acid	20 mL
Activated charcoal	500 mg

2.1 Procedures: Dissolve 1 g of basic fuchsin in 100 mL boiling distilled water. Sufficiently mix the solution and cool to 50°C. Add 2 g of potassium (sodium) metabisulfite and 20 mL of 1 mol/L hydrochloric acid, placed under the conditions of protection from light at room temperature. Lastly, add 500 mg of activated charcoal and shake for 1 minute to remove impurities in basic fuchsin. Filter the solution and then put in a bottle. The filtered solution should be clear and chartreuse; store it at 4°C in the dark. It is good for half a year. When the solution turns before use, it should be discarded.

2.2 Test for Schiff reagent: Pour 10 mL of 37% formalin into a watch glass. Then add a few drops of the Schiff reagent to be tested. A good Schiff reagent will rapidly turn a red-purple color. A deteriorating Schiff reagent will give a delayed reaction, and the color produced will be a deep blue-purple.

3. Hydrosulfite solution
10% sodium (potassium) metabisulfite	5 mL
1 mol/L hydrochloric acid	5 mL
Distilled water	90 mL

DOI 10.1515/9783110531398-019

II Tissue processing

1. Fixation: 10% formalin or Carnoy solution.
2. Section: Paraffin sections at 4–5 μm.

III Procedures

1. Dewax the sections by passing through two changes of xylol. Rehydrate the sections by passing through 100%, 95%, 80%, 70% and 50% alcohols. Finally, rinse gently with distilled water (2 minutes).
2. Oxidize the sections in 0.5%–1% periodic acid solution for 2 to 5 minutes.
3. Thoroughly rinse the sections with distilled water.
4. Cover the sections with Schiff reagent for 15 minutes (sections become light pink during this step).
5. Rinse the sections with bisulfite solution for 2 minutes, three changes.
6. Rinse the sections with running tap water for 5 minutes (sections immediately turn dark pink).
7. Rinse the sections with distilled water for 1 minute.
8. Stain the nuclei with hematoxylin to improve contrast.
9. Dehydrate the sections in graded ethanols and clear with xylene. Cover using resinous mounting medium.

IV Results

PAS-positive substances (glycogen, neutral mucopolysaccharide and part of the acid mucopolysaccharide) turn purplish red (Fig. 3.2).

Glycoprotein	reddish
Mast cell	red
Nuclei	blue

PAS glycogen staining of control tissue was negative.

V Control experiment

1. PAS reaction blocked by acetylation: The mixed solution contained 16 mL of acetic anhydride and 24 mL of pyridine process compared section for 1 to 24 hours (at 22°C). The rinsed, compared section and the experimental sections simultaneously carry out PAS reaction.
2. Hydrolyze diastase: The sections are hydrolyzed with 1% diastase solution for 40 minutes at 37°C (or for 60 minutes at room temperature). Also, saliva takes the place of diastase: take five drops of saliva in distilled water diluted 5–10

times for processing section (a drop of 1% glacial acetic acid drop on the tip of the tongue accelerates salivary secretion). Change once every 30 minutes, a total of three times.

VI Notes

1. The intensity of PAS reaction results is dependent to some extent on the treatment time, pH and concentration of the periodic acid and Schiff reagent. As the temperature increases and the reaction prolongs the glycol, groups of uronic acid will be oxidized. If the oxidation time is more than 15 minutes, it may lead to a nonspecific reaction.
2. Possible free aldehyde groups within the organization are likely to produce false positives. Positive control experiments should be done. The one adjacent section without periodic acid oxidation is directly put into the Schiff reagent. If it appears red, it is false positive. In case of necessity, block the free aldehyde groups with sodium borohydride.
3. Normal precautions exercised in handling laboratory reagents should be followed. Dispose of waste observing all local, state, provincial or national regulations.
4. Periodic acid solution is corrosive. In case of contact with eyes, rinse immediately with plenty of water and seek medical advice. Take off immediately all contaminated clothing. Wear suitable protective clothing, gloves and eye/face protection.
5. Schiff reagent is toxic. It is harmful if swallowed and causes burns. It may also cause cancer. In case of contact with eyes, rinse immediately with plenty of water and seek medical advice. Wear suitable protective clothing, gloves and eye/face protection. In case of accident or if you feel unwell, seek medical advice immediately (show the label where possible). It is restricted to professional users. Avoid exposure; obtain special instructions before use.

Experiment 2: Feulgen reaction

I Materials and reagents

1. HCl (1 mol/L)
   Hydrochloric acid   8.5 mL
   Distilled water   91.5 mL
2. Sulfurous acid solution
   10% potassium (or sodium) sulfite   5 mL
   1 mol/L HCl   5 mL
   Distilled water   90 mL
3. Schiff reagent

II Procedures

1. Fixation
    The tissue should be fixed by 10% neutral formalin or Carnoy, paraffin embedded and sectioned.
2. Dewax the sections by passing through two changes of xylol. Rehydrate the sections by passing through 100%, 95%, 70% and 50% alcohols. Finally, rinse gently with distilled water (2 minutes).
3. Hydrolyze the sections by passing through 1 mol/L HCl at 60°C for 10 minutes.
4. Extract sections in 1 mol/L HCl at room temperature (1–2 minutes), then rinse gently with fresh distilled water (1–2 minutes).
5. Stain the sections in Schiff reagent for 30–60 minutes (protection from light).
6. Bleach the sections with sulfurous acid (2 minutes) for three times, and rinse gently with fresh distilled water (5 minutes).
7. Dehydrate the sections with ethanol, clear with xylene and mount in a resinous medium.

III Results

DNA should be stained in a purple-magenta color (Fig. 4.1).

IV Notes

1. In the Feulgen procedure, purines are hydrolyzed from DNA, exposing the C_1-aldehydes of deoxyribose. When further hydrolyzed, the nucleic acid would be converted to histone and nucleotide; thus, the timing of hydrolysis is critical and should be strictly controlled.
2. The pH value of Schiff reagent should be controlled between 3.0 and 4.3.

Experiment 3: Lead phosphate method showing Acid phosphatase (ACP)

I Materials and reagents

1. Acetate barbiturate buffer
Sodium acetate	9.714 g
Sodium barbital	14.714 g
Distilled water	500 mL
2. HCl (1 mol/L)

3. Incubation buffer (prepared before use)

Acetic acid barbiturate buffer	1.7 mL
8.5% NaCl	0.7 mL
0.1 mol/L HCl	3.6 mL
Distilled water	24 mL
3.3% lead nitrate	0.25 mL
3.2% sodium β-glycerophosphate	2.6 mL (drops while shaking)
0.6 mol/L MgSO$_4$ (activator)	0.7 mL

    Adjust pH value (0.1 mol/L HCl) to 4.7–4.8.

II Procedures

1. The rat kidney is fixed by 95% alcohol-acetone for 30 minutes (precooled and fixed in a 4°C refrigerator), and then frozen samples are sectioned in cryostat.
2. Incubate the sections for 60 minutes at 37°C.
3. Rinse the sections with double distilled water for 2–3 minutes, and put into 1–2% ammonium sulfide (newly prepared) for 2 minutes.
4. Rinse the sections with running water for 2 minutes.
5. The sections are dehydrated, cleared and sealed with glycerin gelatin.

III Results

The positive positions (cytoplasm of renal tubule cells) are stained brown or brownish black (Fig. 6.8).

IV Control groups

In the negative control group (add inhibitor 0.01 mol/L NaF or remove the substrates from the incubation buffer), the cytoplasm of renal tubule cells is all pale stained.

V Notes

1. The ACP should be fixed in cold fixative because of its soluble properties.
2. In preparing the incubation buffer, sediments were not allowed to present, and the buffer should be completely transparent.
3. The pH value of incubation buffer must be controlled in the accurate range.

Experiment 4: Tetrazolium salt method showing lactate dehydrogenase (LDH)

I Materials and reagents

Preparations for incubation buffer:

1 mol/L D, L-sodium lactate	0.1 mL
NAD (4 mg/mL)	0.1 mL
0.06 mol/L phosphate buffer (pH 7.0)	0.25 mL
NBT (4 mg/mL)	0.25 mL
PMS or 1-methoxy PMS	1.96 mg
0.1 mol/L KCN	0.1 mL
0.5 mol/L MgCl$_2$	0.1 mL
Distilled water	10 mL

II Procedures

1. Rat skeletal muscle is quickly cut and fixed in 2% precooled paraformaldehyde, rinsed with phosphate-buffered saline (PBS) and then frozen and sectioned in cryostat.
2. Tissue sections are rinsed with distilled water for 3 minutes.
3. Incubated in dark at 37°C for 30 minutes.
4. Rinse with normal saline twice, 8 minutes each time.
5. Dehydrated and sealed with glycerin gelatin.

III Results

The positive skeletal muscle fiber is violet-blue stained. The control group (by remove the substrate) shows negative pale stained skeletal muscle fiber (Fig. 17).

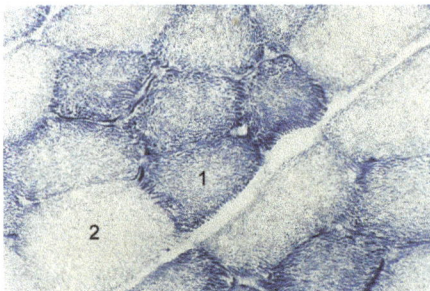

Fig. 17: Tetrazolium salt method shows LDH in skeletal muscle fibers. (1) Positive skeletal muscle fiber. (2) Negative skeletal muscle fiber.

IV Notes

1. If the tissues are not immediately used, it ought to be stored at –20°C.
2. Tissue sections are recommended to rinse with distilled water before treatment. It helps the recovery of the enzyme activities because of low-temperature storage or fixation.

Experiment 5: Tetrazolium salt method showing Succinate dehydrogenase (SDH)

I Materials and reagents

Preparations for incubation buffer:
0.2 mol/L phosphate buffer (pH 7.6)   12 mL
(or Tris-Maleate buffer, pH 7.6, 15.5 mL)
NBT (4 mg/mL)   2.5 mL
0.2 mol/L sodium succinate   12 mL

II Procedures

1. The rat is decapitated, and the skeletal muscle is cut and cryostat sectioned.
2. Frozen sections are fixed by 2% paraformaldehyde for 20 minutes.
3. Rinse the sections with Tris-maleate buffer (pH value is adjusted to 7.4) for 30 minutes, and change the buffer for two times.
4. Incubate the sections for 30 minutes at 37°C, 1–2 hours (the time depended on the reaction conditions, and the recommended time span is 15–35 minutes).
5. Rinse the sections with normal saline for two times, 8 minutes each time.
6. Fix the sections with 10% formalin for 10 minutes, then rinse with water.
7. Dehydrate the sections with ethanol, clear with xylene and mount in a resinous medium.

III Results

The positive skeletal muscle fibers are violet-blue stained. The negative control can be obtained by removing the substrate or by adding malonic acid sodium (final concentration is 3.7 mg/mL), and all the skeletal muscle fibers are pale stained (Fig. 18).

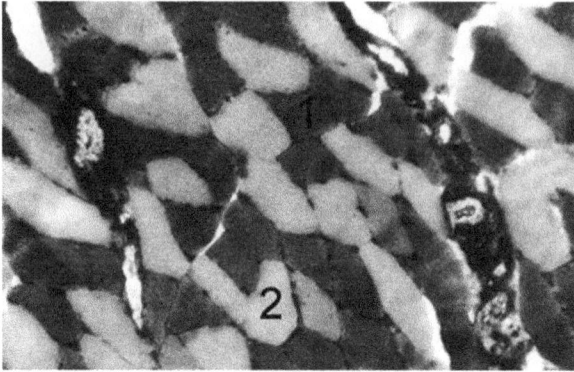

Fig. 18: Tetrazolium salt method shows SDH in skeletal muscle fibers. (1) Positive skeletal muscle fiber. (2) Negative skeletal muscle fiber.

Experiment 6: Immunofluorescence staining method showing human sperm associated antigen 11c

Immunofluorescence histochemical was among the first to set up and one of the widely used immunohistochemical techniques. This method is quick and easy, especially in clinical pathological diagnosis. In addition, not fixed on frozen section is also commonly used immunofluorescence staining.

I Antigen detection method

1.   Materials and reagents
1.1   The primary antibodies are originated from goat polyclonal IgG against human sperm associated antigen 11c.
1.2   The secondary antibody: FITC-labeled donkey antigoat IgG
1.3   PBS (0.1 mol/L, pH 7.2–7.6)
1.3.1   Stock solution A: 0.1 mol/L Na_2HPO_4
        $Na_2HPO_4 \cdot 12H_2O$ 35.8 g
        Distilled water   1000 mL
1.3.2   Stock solution B: 0.1 mol/L NaH_2PO_4
        $NaH_2PO_4 \cdot 2H_2O$ 15.6 g
        Distilled water   1000 mL
1.3.3   Stock solution C: 0.1 mol/L PB (pH 7.2–7.6)
        Stock solution A   81 mL
        Stock solution B   19 mL
        Adjust the pH value to 7.2–7.6 with 1 mol/L HCl or 1 mol/L NaOH

1.3.4   PBS (0.1 mol/L, pH 7.2–7.6)
   Stock solution C 100 mL
   NaCl   `0.9 g

II Procedures

1. The section should be rinsed with 0.01 mol/L, pH 7.4, PBS for 10 minutes and then put horizontally in the wet box.
2. Add the primary antibody at 37°C or room temperature for 30 minutes.
3. Rinse with PBS for 5–10 minutes, twice.
4. Add the secondary antibody labeled with FITC, 37°C for 30 minutes.
5. Rinse with PBS for 5–10 minutes, twice.
6. Seal the section with buffered glycerol and observe under fluorescence microscope.

III Results

The positive structure shows green color (Fig. 8.2).

Experiment 7: PAP immunoenzyme staining

The PAP method is one of the most commonly used immunohistochemical methods. The antienzyme antibodies of PAP complex and the primary antibodies are from the same animal; thus, the secondary antibody can be used as the bridge antibody of combining antibody to connect the PAP complex and the primary antibody, leading the enzyme to the antigen to form the colored final product through enzyme-histochemical reaction.

I PAP method

1. Materials and reagents
1.1 The primary antibody, which originated from different source species: goat anti-A polyclonal IgG
1.2 The secondary antibody: donkey antigoat IgG
1.3 Goat PAP complex
1.4 Blocked serum: normal donkey serum
1.5 The stock solution: 3% H_2O_2
   30% H_2O_2 10 mL
   Distilled water 90 mL

1.6　　Tris-buffered saline (TBS) (0.1 mol/L, pH 7.2–7.6)

1.6.1　Stock solution A: 1 mol/L TB

　　　　Tris　　　　　　　　　60.57 g

　　　　1 mol/L HCl　　　　　210 mL

　　　　Distilled water　　　　Make up to 500 mL

1.6.2　TB (0.1 mol/L, pH 7.6)

　　　　1 mol/L TB　　　　　　10 mL

　　　　Distilled water　　　　90 mL

1.6.3　TBS (0.1 mol/L, pH 7.6)

　　　　1 mol/L　　　　　　　10 mL

　　　　NaCl　　　　　　　　0.9 g

　　　　Distilled water　　　　Make up to 100 mL

　　　　Adjust pH value to 7.6 with 1 mol/L HCl or 1 mol/L NaOH

1.7　　Triton X-100 (1%)

1.7.1　Triton X-100 (30%)

　　　　Triton X-100　　　　　28.2 mL

　　　　0.1 mol/L TBS　　　　72.8 mL

　　　　Mix at 37°C

1.7.2　Triton X-100 (1%)

　　　　30% Triton X-100　　　4 mL

　　　　0.1 mol/L TBS　　　　120 mL

1.8　　3,3'-Diaminobenzidine tetrahydrochloride substrate kit

1.9　　Mounting medium treatment

II Procedures

1. The section paraffin should be removed, and then immerse the sections in distilled water. Frozen or vibration sections can stick onto the glass slides. They can be dried and then rinsed with TBS or dyed by floating in it because TBS can remove the fixative.

2. Immerse the sections in 0.3% H_2O_2 (made with the dual steaming water, TBS or pure methanol) for 10–30 minutes.

3. Rinse the sections twice with TBS, 5–10 minutes each time.

4. Immerse the sections in 0.2%–1% Triton X-100 or 0.1% saponin for 30 minutes.

5. Rinse the sections twice with TBS, 5–10 minutes each time.

6. Drop 3% normal serum, and incubate the sections for 30 minutes; discard the serum without rinsing.

7. Incubate the sections with primary antibody at 4°C for 24–72 hours, and place in the wet box; dye by floating using a 24-hole culture plate sealed with the cover.

8. Rinse the sections twice with TBS, 5–10 minutes each time.

9. Incubate the sections in secondary antibody (such as goat rabbit IgG resistance), 1:100 at the room temperature for 30 minutes.
10. Rinse the sections twice with TBS, 5–10 minutes each time.
11. Incubate the sections in goat PAP complex 1:100 at the room temperature for 30 minutes.
12. Rinse the sections twice with TBS, 5–10 minutes each time.
13. Colorate the sections with 0.05% DAB–0.01% H_2O_2 (prepared with 0.1 mol/L pH 7.6 TB) for 2–5 minutes; control the staining degree with a microscope. Other electron donor is also practical.
14. Rinse the sections with TBS and DDW.
15. Dehydrate the sections with increasing gradient ethanol, clear with xylene and mount in a resinous medium.

III Results

The positive structure shows brown color (Fig. 10.2).

Experiment 8: Streptavidin-ABC (SABC) immunoenzyme staining

The combining sites of the tissue nonspecific avidin also need to be noted. Because avidin contains sugar residues (including mannose and glucosamine), these residues can react with the lectinoid protein in the tissues or adhere to certain tissue components. Apart from this, it has high isoelectric point (pI; 10.5); hence, it can combine with the tissue parts with negative electrons to create nonspecific staining. Streptavidin can be used to replace avidin, namely, the SABC method. Streptavidin is abstracted from *Streptomyces*, but it has no sugar residues, and the pI is close to neutral (6.5). Thus, this method is widely applied at present.

I SABC method

1.   Materials and reagents
1.1 The primary antibody that originated from different source species: goat anti-human polyclonal IgG
1.2 The secondary antibody: biotin-conjugated donkey antigoat IgG
1.3 SABC complex
1.4 Blocked serum: normal donkey serum
1.5 Stock solution: 3% H_2O_2

30% H_2O_2	10 mL
Distilled water	90 mL

1.6 TBS (0.1 mol/L, pH 7.2–7.6)
1.6.1 Stock solution A: 1 mol/L TB
Tris 60.57 g
1 mol/L HCl 210 mL
Distilled water Make up to 500 mL
1.6.2 TB (0.1 mol/L, pH 7.6)
1 mol/L TB 10 mL
Distilled water 90 mL
1.6.3 TBS (0.1 mol/L, pH 7.6)
1 mol/L 10 mL
NaCl 0.9 g
Distilled water Make up to 100 mL
Debugging pH value to 7.6 with 1 mol/L HCl or 1 mol/L NaOH
1.7 Triton X-100 (1%)
1.7.1 Triton X-100 (30%)
Triton X-100 28.2 mL
0.1 mol/L TBS 72.8 mL
Mixed at 37°C
1.7.2 Triton X-100 (1%)
30% Triton X-100 4 mL
0.1 mol/L TBS 120 mL
1.8 3,3'-Diaminobenzidine tetrahydrochloride substrate kit
1.9 Permount mounting medium

II Procedures

1. The paraffin should be removed from the paraffin sections and then immersed in distilled water. Frozen or vibration sections can stick onto the glass slides. They can be dried and then rinsed with TBS or dyed by floating in it because TBS can remove the fixative.
2. Immerse the sections in 0.3% H_2O_2 (made with the dual steaming water, TBS or pure methanol) for 10–30 minutes.
3. Rinse the sections twice with TBS, 5–10 minutes each time.
4. Immerse the sections in 0.2%–1% Triton X-100 or 0.1% saponin for 30 minutes.
5. Rinse the sections twice with TBS, 5–10 minutes each time.
6. Drop 3% normal serum, and incubate the sections for 30 minutes; discard the serum without rinsing.
7. Incubate the primary antibody (high dilution is acceptable) at the room temperature for 30 minutes or 4°C overnight.
8. Rinse the sections twice with TBS, 5–10 minutes each time.

9. Incubate the sections in secondary antibody 1:200 at the room temperature for 30 minutes.
10. Rinse the sections twice with TBS, 5–10 minutes each time.
11. Incubate the SABC compound (1:100) at the room temperature for 30–60 minutes.
12. Rinse the sections twice with TBS, 5–10 minutes each time.
13. Colorate the sections with 0.05% DAB–0.01% H_2O_2 (prepared with 0.1 mol/L pH 7.6 TB) for 1–5 minutes; control the staining degree with a microscope. Other electron donor is also practical.
14. Rinse the sections with TBS and DDW.
15. Dehydrate the sections with increasing gradient ethanol, clear with xylene and mount in a resinous medium.

III Results

The positive structure shows brown color.

Experiment 9: SP double immunoenzyme staining

I Materials and reagents

1. Two kinds of primary antibodies that originated from different source species
1.1 Goat anti-USP26 polyclonal IgG
1.2 Rabbit anti-AR polyclonal IgG
2. Two kinds of secondary antibodies
2.1 Biotin-conjugated donkey antigoat IgG
2.2 Biotin-conjugated donkey antirabbit IgG
3. Blocked serum: normal donkey serum
4. H_2O_2 (3%)
   30% H_2O_2 · 　　　　10 mL
   Distilled water　　　90 mL
5. PBS (0.1 mol/L, pH 7.2–7.6)
5.1 Stock solution A: 0.1 mol/L Na_2HPO_4
   $Na_2HPO_4 \cdot 12H_2O$　　35.8 g
   Distilled water　　　1000 mL
5.2 Stock solution B: 0.1 mol/L NaH_2PO_4
   $NaH_2PO_4 \cdot 2H_2O$　　15.6 g
   Distilled water　　　1000 mL
5.3 Stock solution C: 0.1 mol/L PB (pH 7.2–7.6)
   Stock solution A　　81 mL
   Stock solution B　　19 mL
   Debugging pH value to 7.2–7.6 with 1 mol/L HCl or 1 mol/L NaOH

5.4   PBS (0.1 mol/L, pH 7.2–7.6)
      Stock solution C   100 mL
      NaCl   0.9 g
6.   Triton X-100 (1%)
6.1   Triton X-100 (30%)
      Triton X-100   28.2 mL
      0.01 mol/L PBS   72.8 mL
      Mixed at 37°C
6.2   Triton X-100 (1%)
      30% Triton X-100   4 mL
      0.01 mol/L PBS   120 mL
7.   HRP-labeled streptavidin
8.   AP-labeled streptavidin
9.   3,3′-Diaminobenzidine tetrahydrochloride substrate kit
10.   nitroblue tetrazolium (NBT)/5-bromine-4-chlorine-3-3indole phosphate (BCIP) substrate kit
11.   Permount TM mounting medium

II Procedures

1. Dewax the sections by passing through two changes of xylol. Rehydrate the sections by passing through 100%, 95%, 80%, 70% and 50% alcohols. Finally, rinse gently with distilled water (10 minutes).
2. Inhibit endogenous peroxide activity in the sections with 3% hydrogen peroxide for 20 minutes, and rinse four times with 0.01 mol/L PBS (5 minutes/rinse).
3. Increase the penetration of primary antibodies by rinsing in 1% Triton X-100 for 15 minutes.
4. Rinse the sections gently with running water for 20 minutes, and then rinse two times with 0.01 mol/L PBS (5 minutes/rinse).
5. Block the nonspecific binding sites with normal donkey serum for 25 minutes.
6. Incubate the sections overnight at 4°C in goat anti-A polyclonal IgG.
7. Rinse three times with 0.01 mol/L PBS (5 minutes/rinse).
8. Incubate the sections in biotin-conjugated donkey antigoat IgG for 60 minutes at 37°C, and rinse three times with 0.01 mol/L PBS (5 minutes/rinse).
9. Incubate the sections for 25 minutes in AP-labeled streptavidin, and rinse three times with 0.01 mol/L PBS (5 minutes/ rinse).
10. Display AP: Develop 2–5 minutes in 100 μL BCIP/NBT substrate solution.
11. Rinse three times with 0.01 mol/L PBS (5 minutes/ rinse).
12. Block the nonspecific binding sites with normal donkey serum for 25 minutes.
13. Incubate overnight at 4°C in rabbit anti-B polyclonal IgG.
14. Rinse the sections three times with 0.01 mol/L PBS (5 minutes/ rinse).

15. Incubate biotin-conjugated donkey antirabbit IgG for 60 minutes at 37°C, and rinse three times in 0.01 mol/L PBS (5 minutes/ rinse).
16. Incubate the sections for 25 minutes in HRP-labeled streptavidin, and rinse three times with 0.01 mol/L PBS (5 minutes/ rinse).
17. Display HRP: Develop 10–15 minutes in 100 μL DAB substrate solution.
18. Rinse the sections three times with 0.01 mol/L PBS (5 minutes/rinse).
19. Dehydrate the sections with increasing gradient ethanol, clear with xylene and mount in a resinous medium.

III Results

(A) Antigen should be stained in blue. (B) Positive reaction appears as blue (Fig. 19).

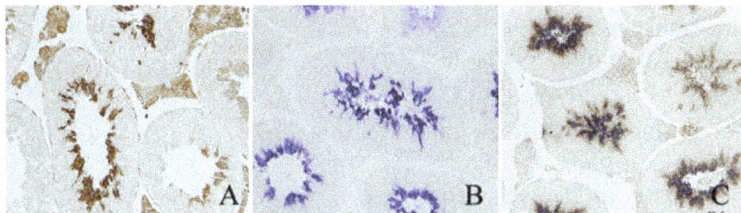

Fig. 19: (A) USP26 protein locates in the testes. (B) AR protein locates in the testes. (C) USP26 and AR were jointly located in the rat testes.

Experiment 10: *In situ* hybridization of prostate-specific antigen mRNA

I Materials and reagents

1. 10× PBS stock solution

NaCl	40.0 g
KCl	1.0 g
KH_2PO_4	1.2 g
Na_2HPO_4	7.2 g
Distilled water	400 mL, adjust pH to 7.4 with NaOH
Distilled water	Make up to 500 mL. Autoclave and store at room temperature.

2. Gelatinum-chromium potassium sulfate sticking reagent

Gelatinum	2.5 g
0.1% DEPC	500 mL, heat to 70°C to dissolve gelatinum. After cooling, add 2.5 g $CrK(SO_4)_2$ and stir into solution.

3. Poly-L-lysine sticking reagent (0.1%)
    Poly-L-lysine   100 mg
    0.1% DEPC-treated H$_2$O   100 mL
4. 0.1% (v/v) DEPC-treated H$_2$O
    DEPC   1 mL
    Distilled water   999 mL
    Allow the solution to stand at 37°C overnight, and autoclave to remove DEPC.
5. Protease K stock solution (1.5%)
    Protease K   15 mg
    100 mmol/L Tris-HCl (pH 8.0)   1 mL
6. Protease K working solution (1 µg/mL)
    100 mmol/L CaCl$_2$ 0.6 mL
    1 mol/L Tris-HCl (pH 8.0)   0.3 mL
    Distilled water   Make up to 30.0 mL
    1.5% Protease K   2 µL
7. Acetic anhydride (make fresh)
    0.1 mol/L trolamine (pH 8.0)   50 mL
    Acetic anhydride   125 µL
8. Prehybridization buffer and hybridization buffer (Fig. 13.3)
9. Hybridization stock solution A
    Na$_2$HPO$_4$ 0.352 g (0.25 mol/L)
    NaCl   1.736 g (3 mol/L)
    EDTA (0.5 mol/L)   500 µL (25 mmol/L)
    Distilled water   Make up to 10 mL
10. Hybridization stock solution B (0.4 mol/L or 6.17% DTT solution)
    Dithiothreitol (DTT)   61.7 mg
    Distilled water   Make up to 1.0 mL

Note: This solution is only needed in case of ^{35}S-labeled probes. Equal volume of distilled water is used instead of this solution in case of non ^{35}S-labeled probes.

11. Hybridization stock solution C (100× Denhardt's solution)
    Bovine serum albumin (BSA)   2.0 g
    Ficoll   2.0 g
    Polyvinylpyrrolidone   2.0 g
    Distilled water   Make up to 100 mL
12. Hybridization stock solution D (2% sodium pyrophosphate solution)
    Sodium pyrophosphate (Na$_4$P$_2$O$_7$)   2.0 g
    Distilled water   Make up to 10 mL
13. Hybridization stock solution E (1% nucleotide solution)
    Sheared salmon sperm DNA   10 mg
    Escherichia coli tRNA   10 mg
    Distilled water   Make up to 1 mL

14. Hybridization stock solution F

Deionized formamide solution: 100 mL formamide + ion-exchange resin (such as Bio-rad AG501-X8, 20–50 mesh), stir for 30 minutes at room temperature and filter. Repeat twice. Dispense into aliquots and store at –20°C.

15. Eluants (Fig. 13.4)
16. 20× SSC

NaCl	175.3 g
Citrate sodium	88.2 g
Distilled water	Make up to 1000 mL

Adjust the pH to 7.04 with 1 mol/L HCl. Dispense into aliquots and sterilize by autoclaving.

II Procedures

1. Fix the prostate tissues in ice-cold 4% paraformaldehyde at 4°C. The fixation should be less than 24 hours to prevent loss of mRNA. Rinse the tissues twice for 30 minutes with 1× PBS.
2. Routine dehydration, clearing, paraffin embedding and sectioning (4–5 μm thick). Bake the sections at 37°C for 1–3 days.
3. Prehybridization pretreatment
3.1 Dewax the sections by treating with xylene and a graded series of ethanol solutions.
3.2 Treat the sections with 0.2 mol/L HCl for 20 minutes at room temperature. Rinse the sections with distilled water for 2 minutes.
3.3 Treat the sections with protease K (1 μg/mL) for 30 minutes at 37°C. Rinse the sections with distilled water for 2 minutes.
3.4 Postfix the sections with 4% paraformaldehyde for 4 minutes at room temperature. Rinse the sections twice for 2 minutes with 1× PBS, then with distilled water for 2 minutes.
3.5 Incubate the sections in freshly mixed 0.25% (v/v) acetic anhydride for 15 minutes at room temperature. Rinse the sections with distilled water for 2 minutes.
3.6 Dehydrate the sections in a graded series of ethanol solutions (increasing concentrations of ethanol) and air-drying.
4. Prehybridization
4.1 Make fresh prehybridization buffer
4.2 Place the sections on a baking plate. Add 20 μL prehybridization buffer onto the tissue.
4.3 Incubate the sections in a humid chamber for 2 hours at 40°C.

5.   Hybridization

5.1   Remove prehybridization buffer. Rinse the sections with DEPC-treated H_2O for 1 minute.

5.2   Immerse in 100% ethanol for 1 minute and air-dry.

5.3   Encircle the tissue with a wax pencil.

5.4   Prepare the hybridization buffer

5.4.1   Mix DNA probe with hybridization stock solutions E and F at a final probe-specific radioactivity of 1×10^6 cpm each section.

5.4.2   Place the sections on a hot plate at 95°C for 10 minutes to unwind DNA. Cool down immediately on ice.

5.4.3   Hybridization stock solution A, B, C and D are added. Mix the solutions.

5.5   Pipette 20 μL hybridization buffer onto the section. Cover with an RNase-free coverslip. Use the rubber cement to seal the coverslip.

5.6   Incubate the sections in a humid chamber for 16 hours at 40°C.

6.   Posthybridization treatments

6.1   Remove the coverslip with a needle. Briefly rinse the sections with eluant A.

6.2   Rinse the sections with eluant A for 1 hour at 50°C.

6.3   Rinse the sections with eluant B for 1 hour at 37°C.

6.4   Rinse the sections with eluant C at room temperature overnight, and stir lightly on a magnetic stirrer.

6.5   Dehydrate the sections in a graded series of ethanol solutions (increasing concentrations of ethanol). Perform radioautography, or store the sections at 4°C.

7.   Autoradiography

7.1   Place the nuclear emulsion in 45°C water bath for 20 minutes.

7.2   Place the melted nuclear emulsion in stand-style slide jar; water bath at 45°C.

7.3   Place the sections in the nuclear emulsion twice. Knock the slides to remove superfluous emulsion.

7.4   Place the sections in the box away from light. Dry the sections slowly for 4 h (place wet papers in the box).

7.5   Place the sections in the slide magazine (with drying agent), which is enveloped in a piece of black paper. Allow exposure take place at 4°C for proper time.

7.6   Restore the slide magazine to room temperature. The prepared developer D19, fixer Kodak F-5 and distilled water cool down to 14°C in freezer.

7.7   Develop an image for 4 minutes. Rinse with water for 10 seconds. Fix for 4 minutes. Rinse with water for 20 minutes.

7.8   Take the sections out of dark room. Clean the slide backside and areas surrounding the tissue.

7.9   Contrast stain with hematoxylin, Giemsa or methyl green.

7.10   Routine dehydration, clearing and mounting.

III Results

Silver grains in the positive cells are obviously more numerous than background. The silver grains are uniform in size and well distributed mainly in cytoplasm (Fig. 20).

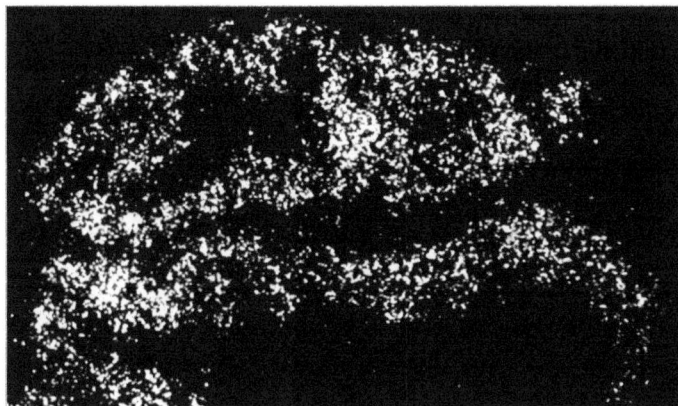

Fig. 20: PSA mRNAs are detected by 35^S-labeled cDNA probes by hybridization. Dark field observation.

Experiment 11: *In situ* hybridization for protein kinase C βI mRNA

I Materials and reagents

1.   PKC βI *In situ* hybridization detection kit containing the following reagents:
1.1   Pepsin (10×)
1.2   Prehybridization solution
1.3   PKC βI oligonucleotide probe hybridization solution
1.4   Blocking solution
1.5   Biotinylated secondary antibody to digoxigenin
1.6   SABC-POD
1.7   Biotinylated peroxidase
2.   DAB color-developing kit
3.   DEPC-treated H_2O (0.1%, v/v)
         DEPC   1 mL
         Distilled water   999 mL

Allow the solution to stand at 37°C overnight; autoclave to remove DEPC.
4.   Poly-L-lysine sticking reagent (0.1%)
         Poly-L-lysine   100 mg
         0.1% DEPC-treated H_2O 100 mL

5. Citric acid (3%)
    Citric acid   3 g
    0.1% DEPC-treated H_2O 100 mL
    Adjust the pH to 2.0
6. 2× SSC
    Trisodium citrate   8.8 g
    NaCl   17.6 g
    0.1% DEPC-treated H_2O 1000 mL
7. 0.5× SSC
    2× SSC   100 mL
    0.1% DEPC-treated H_2O 300 mL
8. 0.2× SSC
    2× SSC   30 mL
    0.1% DEPC-treated H_2O 270 mL
9. Glycerol (20%)
    Glycerol   20 mL
    0.1% DEPC-treated H_2O 80 mL
10. PBS
    $NaH_2PO_4 \cdot 2H_2O$ 0.4 g
    $Na_2HPO4 \cdot 12H_2O$ 6.0 g
    NaCl   30 g
    0.1% DEPC-treated H_2O 1000 mL
11. Paraformaldehyde (4%)/PB (0.1 mol/L) (pH 7.4)
    Paraformaldehyde   40 g
    $NaH_2PO_4 \cdot 2H_2O$ 2.964 g
    $Na_2HPO4 \cdot 12H_2O$ 28.998 g
    0.1% DEPC-treated H_2O 1000 mL
    Adjust the pH to 7.2–7.6.

II Procedures

1. Perform perfusion fixation of the rats. Postfix the tissue for 30 minutes. Perform routine dehydration, clearing and paraffin embedding.
2. Perform continuous corona sectioning (8 µm thick). Gain the sections with intact hippocampi structure. Bake the sections at high temperature for 4 h, then at 37°C for 3 days.
3. Dewax sections by treating with xylene and a graded series of ethanol solutions.
4. Treat the sections for 10 minutes with 3% H_2O_2. Rinse the sections two times with distilled water.
5. Pipette the diluted pepsin solution (add two drops of concentrated pepsin to 1 mL 3% citric acid) onto the sections to expose the mRNA fragments. Treat the sections for 15 minutes at 37°C.

6. Rinse the sections thrice for 5 minutes with PBS buffer. Rinse the sections with distilled water.
7. Prehybridization treatment: Pipette 20 µL of prehybridization solution onto each section. Incubate the sections in a humid chamber (20% glycerol is added to the bottom of the humid chamber to maintain humidity) for 4 hours at 37°C–40°C. Draw the redundant solution. Do not rinse the sections.
8. Hybridization: Pipette 20 µL of hybridization solution onto each section. Cover with a coverslip. Incubate the sections at 37°C–40°C overnight.
9. Posthybridization rinse: Remove the coverslip. Rinse the sections twice for 5 minutes at 30°C–37°C with 2× SSC, twice for 15 minutes with 0.5× SSC and twice for 15 minutes with 0.2× SSC.
10. Pipette the blocking solution onto the sections. Incubate the sections at 37°C for 30 minutes. Draw the redundant solution. Do not rinse the sections.
11. Pipette the biotinylated secondary antibody to digoxigenin onto the sections. Incubate the sections at 37°C for 60 minutes. Rinse the sections four times for 5 minutes with PBS buffer.
12. Pipette the SABC onto the sections. Incubate the sections at room temperature for 30 minutes. Rinse the sections thrice for 5 minutes with PBS buffer.
13. Pipette the biotinylated peroxidase onto the sections. Incubate the sections at room temperature for 30 minutes. Rinse the sections four times for 5 minutes with PBS buffer.
14. DAB color developing: Prepare DAB color reagent as recommended by the manufacturer. (Mix a drop of color-developing reagent A, a drop of color-developing reagent B and a drop of color-developing reagent C in 1 mL distilled water.) Add 50 µL 8% $NiCl_2$ for every 50 mL color-developing reagent. Pipette the color reagent onto the sections. Incubate the sections at room temperature for 8 minutes. Stop the color reaction by rinsing the sections several times with tap water.
15. Perform routine dehydration, clearing and mounting.

III Results

The positive staining appears purple in color (Fig. 21).

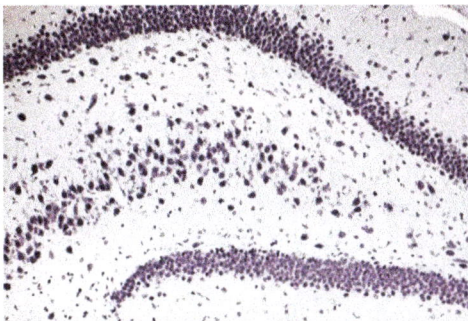

Fig. 21: Biotin-labeled probes are introduced for *in situ* hybridization for PKC βI mRNA and are detected by ABC method and visualized by nickel-DAB-H_2O_2.

Experiment 12: *In situ* hybridization of testis expressed gene 11 mRNA

I Materials and reagents

1. HCL (200 mmol/L)
Concentrated HCl	820 µL
0.1% DEPC-treated H_2O	50 mL
2. Acetic anhydride (0.5%) (make fresh)
Acetic anhydride	250 µL
1 mol/L Tris-HCl (pH 8.0)	5 mL
0.1% DEPC-treated H_2O	44.75 mL
3. Proteinase K working solution (20 µg/mL)
1 mol/L Tris-HCl (pH 8.0)	5 mL
Proteinase K (10 mg/mL)	100 µL
0.5 mol/L EDTA (pH 8.0)	5 mL
0.1% DEPC-treated H_2O	40 mL
4. EDTA (0.5 mol/L, pH 8.0)
EDTA-Na·$2H_2O$	186.1 g
0.1% DEPC-treated H_2O	800 mL

Stir vigorously on a magnetic stirrer. Adjust the pH to 8.0 with NaOH (approximately 20 g NaOH). Adjust the volume to 1000 mL. Dispense into aliquots and sterilize by autoclaving.

5. Hybridization buffer
20× SSC	50 µL (2× SSC)
50% dextran sulfate	100 µL (10%)
10 mg/mL sheared salmon sperm DNA	5 µL (0.01%)
1% sodium dodecyl sulfate (SDS)	10 µL (0.02%)
100% deionized formamide	250 µL (50%)
0.1% DEPC-treated H_2O	85 µL
6. SDS (10%)
Electrophoresis-grade SDS	100 g
0.1% DEPC-treated H_2O	Make up to 1000 mL

Heat to 68°C to assist dissolution. If necessary, adjust the pH to 7.2 by a few drops of concentrated HCl. Store at room temperature. Do not autoclave.

7. Blocking solution
10× Blocking stock solution (10%)	5 mL
Fetal bovine serum	5 mL (10%)
Maleic acid buffer	40 mL
8. 10× Blocking stock solution (10%)
Blocking reagent for nucleic acid hybridization	10 g
Maleic acid buffer	100 mL

Heat to assist dissolution. Sterilize by autoclaving. Dispense into aliquots and store at –20°C.

9. Maleic acid buffer (100 mmol/L maleic acid; 150 mmol/L NaCl, pH 7.5)

1 mol/L maleic acid	5 mL
5 mol/L NaCl	1.5 mL
0.1% DEPC-treated H_2O	Make up to 50.0 mL

Adjust the pH to 7.5 with NaOH. Sterilize by autoclaving.

10. Maleic acid (1 mol/L)

Maleic acid	34.83 g
0.1% DEPC-treated H_2O	Make up to 300 mL

Sterilize by autoclaving

11. TBS buffer

1 mol/L Tris-HCl (pH 7.5)	25 mL
5 mol/L NaCl	15 mL
0.1% DEPC-treated H_2O	460 mL

12. Tris-HCl (1 mol/L): Dissolve 121.1 g Tris base in 800 mL 0.1% DEPC-treated H_2O. Adjust the pH to the desired value by concentrated HCl.

pH	HCl
7.4	70 mL
7.6	60 mL
8.0	42 mL

Allow the solution to cool to room temperature before making final adjustments to the pH. Adjust the volume of the solution to 1 L with 0.1% DEPC-treated H_2O. Dispense into aliquots and sterilize by autoclaving.

13. NaCl (5 mol/L)

NaCl	292.2 g
0.1% DEPC-treated H_2O	Make up to 1000 mL

Dispense into aliquots and sterilize by autoclaving.

14. Detection buffer (100 mmol/L Tris-HCl, 100 mmol/L NaCl, pH 9.5)

1 mol/L Tris-HCl (pH 9.5)	5 mL
5 mol/L NaCl	1 mL
Distilled water	Make up to 50 mL

II Procedures

1. Tissue preparation

Fix the testes in ice-cold 4% paraformaldehyde at 4°C. Perform routine dehydration, clearing, paraffin embedding and sectioning.

2. Prehybridization treatment

2.1 Dewax the sections by treating with xylene and a graded series of ethanol solutions.

2.2   Fix the sections once again with 4% paraformaldehyde (in 100 mmol/L phosphate buffer) for 20 minutes. Rinse the sections three to five times with TBS buffer.

2.3   Treat the sections for 10 minutes with 200 mmol/L HCl to denature proteins.

2.4   Incubate the sections for 10 minutes on a magnetic stirrer with a freshly mixed solution of 0.5% acetic anhydride in 100 mmol/L Tris (pH 8.0). Rinse the sections three to five times with TBS buffer.

2.5   Treat the sections for 20 minutes at 37°C with proteinase K (10–500 μg/mL in TBS, which contains 2 mmol/L CaCl$_2$). The concentration of proteinase K depends on the degree of fixation. Generally, start with 20 μg/mL proteinase K for paraformaldehyde-fixed tissue. Rinse the sections three to five times with TBS buffer.

2.6   Incubate the sections at 4°C for 5 minutes with TBS (pH 7.5) to stop the proteinase K digestion.

2.7   Dehydrate the sections in a graded series of ethanol solutions (increasing concentrations of ethanol).

2.8   Rinse the sections briefly with chloroform.

3.   Hybridization

3.1   Place the sections in a humid chamber at 55°C for 30 minutes.

3.2   Prepare a hybridization buffer

3.3   Dilute the digoxigenin (DIG)-labeled antisense RNA probe in hybridization buffer (working concentration of antisense RNA probe is 0.2–10 ng/μL).

3.4   Pipette the diluted antisense RNA probe solution onto each section at a volume of 10 μL/cm^2. Cover with a coverslip. Use the rubber cement to seal the coverslip.

3.5   As a control, serve sections treated in the same way whereby in steps 3 and 4 a labeled sense RNA probe is used.

3.6   Place the sections on a hot plate at 95°C for 4 minutes.

3.7   Incubate the sections in a humid chamber for 4–6 hours at 55°C–75°C.

4.   Posthybridization rinse

4.1   Remove the rubber cement. Incubate the sections in 2× SSC overnight.

4.2   Incubate thrice for 20 minutes at 55°C with 50% deionized formamide in 1× SSC.

4.3   Incubate twice for 15 minutes at room temperature with 1× SSC.

4.4   Rinse the sections three to five times with TBS.

5.   Detection of mRNA

5.1   Incubate the sections for 15 minutes with blocking mixture.

5.2   Incubate the sections for 60 minutes with alkaline-phosphatase-conjugated anti-digoxin antibody (diluted 1:500 in blocking mixture).

5.3   Rinse the sections three to five times with TBS.

5.4   Prepare NBT/BCIP color reagent as recommended by the manufacturer. (Mix 45 μL NBT solution and 35 μL BCIP solution in 10 mL of detection buffer.)

5.5   In a Coplin jar in the refrigerator, incubate the sections with color reagent until sufficient color develops.

5.6   Stop the color reaction by rinsing the sections several times with tap water.

5.7   Rinse the sections with distilled water.

5.8   Mount the sections with water-soluble mounting medium (glycerol vinyl alcohol (GVA) mount).

III Results

The positive staining appears black-blue in color (Fig. 13.3).

Index